Optimization Methods in Operations Research and Systems

Optimization Methods in Operations Research and Systems

Contributors

Zhuhong Zhang, Min Liao et al.

AURIS
Reference

www.aurisreference.com

Optimization Methods in Operations Research and Systems

Contributors: Zhuhong Zhang, Min Liao et al.

Published by Auris Reference Limited

www.aurisreference.com

United Kingdom

Optimization Methods in Operations Research and Systems

ISBN: 978-1-78154-711-3

British Library Cataloguing in Publication Data
A CIP record for this book is available from the British Library

Printed in the United Kingdom

Exclusively distributed by CBS Publishers & Distributors Pvt. Ltd.

Sales & Distribution Rights only for India, Pakistan, Bangladesh, Sri Lanka, Nepal and Bhutan.This book is not to be sold outside these territories.

Contents

List of Abbreviations

CD	Coverage Density
DCMMO	Dynamic Constrained Multi-Objective Multimodal Optimization
DMO	Dynamic Multi-Objective Optimization
DPVM	Dynamic Pressure Vessel Model
DSR	Dynamic Speed Reducer Model
CVRP	Capacitated Vehicle Routing Problem
VRP	Vehicle Routing Problem
VRPTW	Vehicle Routing Problem with Time Windows
ANP	Analytical Network Process
GICO	Green INVESTMENT COST OPTIMIZATION
GSCM	Green Supply Chain Management
OECD	Organization for Economic Cooperation and Development
MVS	Monotonic Vector Space
MVRL	Monotonic Vector Requirement Locus
MEH	Minimal Envelop Hyperbox
RV	Representative Value
CPB	Capability-Based Planning
FMIRL	Fuzzy Monotonic Indicator Requirement Locus
DEA	Data Envelopment Analysis
AHP	Analytic Hierarchy Process
DMU	Decision Making Unit
CSW	Common Set of Weights
UNDP	United Nations Development Programme
HPI	Human Poverty Index
HDI	Human Development Index
GDP	Gross Domestic Product
DCF	Discounted Cash Flow
NPV	Net Present Value
HOL	Head-of-Line
QOS	Quality of Service
JPGF	Joint Probability Generating Function
PQS	Priority Queuing Systems
DO	Dynamic Optimization
KKT	Karush-Kuhn-Tucker
VOP	Vector Optimization Problems
IS	Information Systems
IMRT	Intensity Modulated Radiation Therapy
MILP	Mixed-Integer Linear Programming
DVH	Dose-Volume Histograms

List of Contributors

Zhuhong Zhang
Institute of System Science & Information Technology, College of Science, Guizhou University, Guiyang, China

Min Liao
Institute of System Science & Information Technology, College of Science, Guizhou University, Guiyang, China

Lei Wang
Institute of System Science & Information Technology, College of Science, Guizhou University, Guiyang, China

Mariano Frutos
Department of Engineering, National University of the South, Bahía Blanca, Argentina

Fernando Tohmé
Department of Economics, National University of the South, Bahía Blanca, Argentina

Seungbae Sim
Center for Defense Acquisition, Korea Institute for Defense Analyses, Seoul, Republic of Korea

Hosang Jung
Graduate School of Logistics, Inha University, Incheon, Republic of Korea

Jianwen Hu
National University of Defense Technology, Changsha, China

Mohammad Sadegh Pakkar
Faculty of Management, Laurentian University, Sudbury, Canada

Samuel Ambapour
Institute National de la Statistique, Brazzaville, Republic of Congo

Hualun Zhang
School of Public Affairs, University of Science and Technology of China, Hefei, China

Wei Song
School of Public Affairs, University of Science and Technology of China, Hefei, China

Xiaobao Peng
School of Public Affairs, University of Science and Technology of China, Hefei, China

Xiaoyan Song
School of Public Affairs, University of Science and Technology of China, Hefei, China

Hui-Chuan Chen
University of Texas at Arlington, US

Faouzi Kamoun
College of Technological Innovation, Zayed University, Dubai, UAE

Chuma R. Nwozo
Department of Mathematics, University of Ibadan, Ibadan, Nigeria

Charles I. Nkeki
Department of Mathematics, University of Benin, Benin City, Nigeria

Surjeet Kaur Suneja
Department of Mathematics, Miranda House College, University of Delhi, Delhi, India

Sunila Sharma
Department of Mathematics, Miranda House College, University of Delhi, Delhi, India

Meetu B. Grover
Department of Mathematics, Miranda House College, University of Delhi, Delhi, India

Malti Kapoor
Department of Mathematics, Moti Lal Nehru College (M), University of Delhi, Delhi, India

Andrey Kostogryzov
Research Institute of Applied Mathematics and Certification, Moscow, Russia

Leonid Grigoriev
Gubkin Russian State University of Oil and Gas, Moscow, Russia

George Nistratov
Research Institute of Applied Mathematics and Certification, Moscow, Russia

Andrey Nistratov
Institute of Informatic Problems of the Russian Academy of Sciences, Moscow, Russia

Vladimir Krylov
Scientific & Technical Center "NAOPRO", Moscow, Russia

Delal Dink
School of Chemical Engineering, Purdue University, W. Lafayette, USA

Mark Langer
Department of Radiation Oncology, Indiana University, Indianapolis, USA

Seza Orcun
e-Enterprise Center, Discovery Park, Purdue University, W. Lafayette, USA

Joseph Pekny
School of Chemical Engineering, Purdue University, W. Lafayette, USA

Ronald Rardin
Department of Industrial Engineering, University of Arkansas, Fayetteville, USA

Gintaras Reklaitis
School of Chemical Engineering, Purdue University, W. Lafayette, USA

Behlul Saka
Department of Industrial Engineering, University of Arkansas, Fayetteville, USA

Preface

Operations research encompasses a wide range of problem-solving techniques and methods applied in the pursuit of improved decision-making and efficiency, such as simulation, mathematical optimization, queuing theory, economic methods, data analysis, statistics, neural networks, expert systems, and decision analysis. The text Optimization Methods in Operations Research and Systems presents the elementary mathematical introduction to linear and non-linear programming, dynamic programming and geometric programming. First chapter investigates one immune optimization approach for dynamic constrained multi-objective multimodal optimization in terms of biological immune inspirations and the concept of constraint dominance. Second chapter presents a new approach to the analysis of complex distribution problems under capacity constraints. The objective of third chapter is to develop a model that determines the optimal points for investment in green management by defining a mathematical relationship between carbon trading profits and investments in green management using a company's supply chain information. Concepts and operations of monotonic vector space model (I) have been presented in fourth chapter. In fifth chapter, the optimization model is used to get the optimized expectation values of the statistical capability indicators according to practical problems. Sixth chapter proposes an integrated approach to the data envelopment analysis (DEA) and analytic hierarchy process (AHP) methodologies for ratio analysis. Seventh chapter discusses how to assess the relative efficiency of health systems in Sub-Saharan Africa using data envelopment analysis. Eighth chapter applies the data envelopment analysis (DEA) to Chinese provincial panel data from the year 2003 to 2008 for measuring the investment efficiencies and identifying their trends of Chinese 30 provinces and autonomous regions. Ninth chapter provides an overview of real options theory and investigates the relevant literature to determine the gaps which occur when real options are implemented in the operations management field. In tenth chapter, we consider a tandem of two head-of-line (HOL) non-preemptive priority queuing systems, each with a single server and a deterministic service time. Eleventh chapter examines the allocation of resource to different tasks in a production company. A different approach to cone-convex optimization has been presented in twelfth chapter. The goal of thirteenth chapter is to propose models, methods, and software tools well-tested in practice, to predict and optimize quality and risks as applied to newly developed and currently operated manufacture, power generation, transport, engineering, information, control and measurement, food storage, quality assurance and security systems. Last chapter proposes a new mixed-integer linear programming model to simultaneously incorporate fractionation and cumulative constraints in intensity modulated radiation therapy (IMRT) planning optimization used in cancer treatment.

Chapter 1

IMMUNE OPTIMIZATION APPROACH FOR DYNAMIC CONSTRAINED MULTI-OBJECTIVE MULTIMODAL OPTIMIZATION PROBLEMS

Zhuhong Zhang, Min Liao, Lei Wang

Institute of System Science & Information Technology, College of Science, Guizhou University, Guiyang, China

ABSTRACT

This work investigates one immune optimization approach for dynamic constrained multi-objective multimodal optimization in terms of biological immune inspirations and the concept of constraint dominance. Such approach includes mainly three functional modules, environmental detection, population initialization and immune evolution. The first, inspired by the function of immune surveillance, is designed to detect the change of such kind of problem and to decide the type of a new environment; the second generates an initial population for the current environment, relying upon the result of detection; the last evolves two sub-populations along multiple directions and searches those excellent and diverse candidates. Experimental results show that the proposed approach can adaptively track the environmental change and effectively find the global Pareto-optimal front in each environment.

INTRODUCTION

In real-world engineering problems, a great number of optimization problems often involve in multiple timevarying multimodal sub-objectives and constraints, such as portfolio investment, project planning management, chemical engineering design and transportation. These belong to dynamic constrained multi-objective multimodal optimization (DCMMO) which the sub-objective functions, constraints, dimensions of the objective space and design space may change over time. The major challenge lies in that the global Pareto front may move toward another one with time; it is very difficult for a search procedure to keep the sufficient diversity and convergence, due to multiple local Pareto fronts and constraints. Although some techniques are popular in the context of intelligent optimization, they expose many faults when directly applied to DCMMO problems, e.g., adaptation, genetic diversity and constraint-handling. Dynamic multi-objective optimization (DMO) has become active in the recent years. Many researchers in the fields of evolutionary computation and immune optimization have made their efforts to study how the search procedure can adaptively track the environmental change and rapidly find the desired Pareto front [1-10]. DCMMO is one particular kind of DMO, presenting complex dynamic behaviors such as dynamics and multimodality. In order to solve such special kind of problem, we must solve at least three crucial issues: 1) infeasible individuals toward feasible ones, 2) population diversity, and (3) finding the well-distributed global Pareto front.

Although immune optimization, a popular research branch, was proved to be potential for dynamic problems because of the inherent diversity and adaptation, more studies on it are concentrated on solving non-constrained dynamic single or multi-objective optimization problems [10-12]. It is not clear whether bio-immune inspirations are valuable for handling DCMMO. To our knowledge, it is possible to explore efficient and effective immune techniques for such kind of problem. Thereby, we in the present work try studying an immune-inspired optimization approach, dynamic constrained multi-objective multimodal immune optimization approach (DCMMIOA). The comparative experiments draw the strong conclusion that DCMMIOA is a competitive optimizer capable of rapidly tracking the environmental change and effectively discovering the location of the global Pareto front in each environment.

RELATED WORK SURVEY ON INTELLIGENT OPTIMIZATION FOR DMO

Dynamic Multi-Objective Evolutionary Algorithms

DMO has recently gained great attention among researchers in the area of intelligent optimization. The key of designing an advanced technique for such problem is to consider such crucial factors as computational complexity, environmental adaptation and solution quality. Several researchers [1-7] have reported their excellent achievements suitable for non-constrained DMO problems. For example, Farina, et al. [1] constructed five dynamic multi-objective test problems with slowly changing environments and fixed dimensions, based on the static benchmark test problems; meanwhile, one-directional search-based method was proposed to solve them. This method can obtain some Pareto-optimal solutions with somewhat uniform distribution for a given problem, but time consumption is expensive. In [2], two similar dynamic optimization techniques originating from NSGAII, i.e., DNSGAII-A and DNSGAII-B, were proposed to find a minimum frequency of change allowed in the problem to adequately track the theoretical Pareto front on-line. Their main difference is only with the aspect of generating their respective initial populations. One of their merits is that they can also solve DCMO problems. In the DMO studies done by Zhou and Hatzakis [4,5], some prediction-based inspirations were merged into one reported multi-objective evolutionary algorithm, in which the time serial analysis method was used to predict the location of the Pareto front. In addition, in the work made by Wang [6], a new evolutionary algorithm (DMEA) was proposed to deal with DCMO problems, by improving the operators of crossover and mutation. In a study done by Tan [7], one competitive and cooperative mechanisms-based co-evolutionary multi-objective algorithm was developed to solve multi-objective optimization problems in dynamic environments. For one such algorithm, each species subpopulation competes to represent a particular subcomponent of the multi-objective problem, while the eventual winners are required to coevolve so that those better solutions can be found.

Dynamic Multi-Objective Immune Optimization (DMIO)

Since Carlos et al. proposed a simple artificial immune system to solve static multi-objective optimization problems [13, 14], multi-objective immune optimization has become increasingly active. Correspondingly, many original or improved immune techniques [15-17], based on the humoral immune response principle, have been reported continually. This indicates that the artificial immune system paradigm is an interesting topic for multiobjective optimization problems. However, less work on DMIO is displayed in the literature. Shang et al. [8] suggested one clonal selection algorithm for dynamic nonconstrained multi-objective optimization problems, while such approach was examined by two theoretical test problems. More recently, we also studied immune-based optimization techniques for such kind of problem [9]. Subsequently, we handled one general kind of DCMO by developing an immune agent-based artificial immune system (DCMOAIS) [10]. In this approach, the interactive mechanism between B-cell and T-cell is our major bio-inspiration. The experimental results hint that such approach is a competitive optimizer for complex and high-dimensional DCMO problems. Despite of some reported achievements on DMIO, they might be difficult in solving multimodal optimization problems. Thereby, we in this work design DCMMIOA to solve specially DCMMO problems rather than general DCMO. It should be pointed out that although DCMMIOA and DCMOAIS share some bio-immune inspirations, their modules are completely different. For example, DCMMIOA does not include memory pool, while detecting the environmental change only by means of the joint environment; it carries out evolution on two sub-populations, in which the better subpopulation executes multiple evolution schemes so that some global Pareto-optimal solutions can be found.

PROBLEM FORMULATION AND IMMUNE THEORY

Problem Formulation

Consider the dynamic constrained multi-objective multimodal optimization problem (VP):

$$\min_{x\in\Omega(t)} f(x,t)=\left(f_1(x,t)\cdots f_m(x,t)\right)$$

$$s.t.\begin{cases} g_i(x,t)\le 0, i=1,\cdots,I, \\ \Omega(t)=\prod_{i=1}^{n(t)}\left[a_i(t),b_i(t)\right], \end{cases}$$

with discrete time integer variable t, $1\le t\le T$, and bounded and closed domain $\Omega(t)$ in $R^{n(t)}$, where n(t) is the dimension at the moment t; $f(x,t)$ denotes the time-varying vector-valued objective function with at least one multimodal sub-objective function; $g_i(x,t)$, $1\le i\le I$, are the linear or nonlinear inequality constraint functions. Since t takes integers in the interval [1, T], the above problem consists of a series of jointly static multiobjective multimodal problems; each static problem is called an environment; for example, we call the t-th static problem the t-th environment. Our task is to develop DCMMIOA which can continually track the environmental change and rapidly search the global Pareto front in the t-th environment before the next environment arrives.

Additionally, $x\in\Omega(t)$ is said to be feasible for the t-th environment, if it satisfies all the above constraints, and infeasible otherwise. For a given environment t and $x,y\in\Omega(t)$, we say that x constrained-dominates y if one of three conditions is true [2]: (1) x is feasible but y is not; (2) x and y are feasible and $x\prec y$; (3) x and y are both infeasible, but x has a smaller constraint violation $G(x,t)$. Here, $x\prec y$ denotes that $f_l(x,t)\le f_l(y,t)$, $1\le l\le m$, and there exists j, $1\le j\le m$, such that $f_j(x,t) < f_j(y,t)$; meanwhile, $G(x,t)$ represents the total of constraint violations at the point x for all the constraints; namely, $G(x,t)$ is the sum of $\max\{g_i(x,t),0\}$ with $1\le i\le I$. For a given environment t, $x\in\Omega(t)$ is called a local Pareto-optimal solution, if it is feasible, and there exists a neighborhood in x such that any element in such region does not constrained-dominate x; $x\in\Omega(t)$ is said to be a global Pareto-optimal solution, if it is feasible, and all elements in $\Omega(t)$ do not constrained-dominate x.

Immune Theory

The immune response theory describes essentially a process that B and T cells learn an invader and ultimately eliminate it. T cells as detectors can monitor whether a change takes place in the immune system, being capable of identifying "self" and "non-self". When an organism is attacked by the invader or antigen, T cells will experience three phases: initiation, reaction and elimination; namely, a large number of virgin cells are first generated, and then these cells become effectors which urge B cells to respond to the antigen. This stimulates these B cells to create plasma and memory cells through proliferation. If being active, such plasma cells suffer a process of affinity maturation through somatic maturation, and secrete antibodies neutralizing the triggering invader. Subsequently, high-affinity B cells are selected into the B-cell pool, but others are eliminated. Besides, the memory cells will become long-lived ones. Once the previous invader is found in the immune system, these memory cells commence rapidly differentiating into plasma cells capable of producing high-affinity antibodies.

Summarily, by taking an analogy between the two processes of the immune response and solving DCMMO, it is not difficult to know that the task of such response is to create excellent B cells and the purpose of handling DCMMO is to find those best solutions. From the viewpoint of engineering application, such immune response is a bio-inspiration used in developing immune-inspired optimizers. In this work, two immune metaphors of Tcell surveillance and B-cell learning (cell selection, clonal expansion and somatic mutation) are simply simulated to construct our DCMMIOA. The first is adopted to design the module of environmental detection which detects the environmental change; the second is an important biological idea used to create valuable B cells or individuals.

ALGORITHM FORMULATION AND MODULE ILLUSTRATION

Algorithm Formulation

As associated to problem (VP) as in Section 3, for the t-th environment antigen $Ag(t)$ is viewed as the attribute set of $n(t)$, $f(.,t)$ and $\bar{g}(.,t)$, where $\bar{g}(.,t)$ is the vector-valued function composed of constraint functions; B cells are regarded as real-encoded candidate solutions; memory cells denotes those best B cells found until now (i.e., non-dominated candidates). DCMMIOA includes five modules of environmental detection, initialization, division, mutation and recruitment. Following these descriptions, it can be in detail described as follows.

1. Input: Population size N, total of environments T, elitist subpopulation size N_e, clonal size m_c, and mutation rates p_m.
2. While $t \leq T$ do
3. Generate population A with size N, A :=Initialization($Ag(t)$);
4. While $Ag(t)$ does not change do
5. Divide A into elitist and unselected sub-populations, (S_1, S_2):=Division (A, N_e);
6. Carry out elitist mutation on S_1, B_1 :=Mutation (S_1, N);
7. Enforce uniform mutation on S_2, and acquire population B_2;
8. Pick up N better cells in $A \cup B_1 \cup B_2$ to form population C;
9. Recruit new cells, A := Recruitment (C);
10. Implement environmental detection: Detection $(Ag(t))$;
11. End while
12. End while
13. Output the optima found in each environment.

Notice that in the above approach, steps 4 to 11 is a loop of optimization for the t-th environment. Through step 10, once such environment takes a change, the search process in the environment is required to stop. Steps 6 and 7 formulate two sub-populations to evolve along different directions. Additionally, some optimization problems usually include multiple constraints so that their feasible regions are empty or extremely narrow. In such case, it becomes difficult for any optimization approach to solve them. Thus, when handling this kind of hardconstraint problem, the requirement of feasibility is relaxed. Here, we design a threshold index to decide whether an individual is feasible or not, namely

$$\sigma_t = \min\left\{\frac{\sum_{i=1}^{N} G(x_i,t)}{N \times I}\left(1-\frac{n}{M_n(t)}\right), 0.1\right\} \tag{1}$$

where n and $M_n(t)$ denote the iteration number and the maximal iteration in the t-th environment, respectively. If $G(x,t) < \sigma_t$, we say that x is feasible. The main modules are designed below.

Module Illustration

Detection (Ag(t)). This module detects whether Ag(t) has made a change within the runtime, and decides the relation between Ag(t) and Ag(t + 1) if so. Note that Ag(t) comprises of three factors, i.e., n(t), f(., t) and $\bar{g}(.,t)$. If any change takes place among such three factors, we say that the t-th environment has changed. Suppose that such environment has been converted to environment t + 1; so, we decide whether Ag(t + 1) is new or similar through the following way:

1) If $n(t+1) \neq n(t)$, Ag(t + 1) is regarded as a new antigen;
2) If $n(t+1) = n(t)$, we pick randomly up K cells in the eventual population A acquired in the t-th environment to constitute a sample set S, and then check the relation between Ag(t) and Ag(t + 1) through an identification factor η defined as

$$\gamma = \frac{1}{2K}\sum_{i=1}^{K}\Big[\|f(x_i,t+1)-f(x_i,t)\| + \|\bar{g}(x_i,t+1)-\bar{g}(x_i,t)\|\Big],\ x_i \in S \tag{2}$$

where $\|x\|$ stands for the 2-norm of x. If $\gamma < \tau$, we say that Ag(t + 1) is similar to Ag(t); otherwise, Ag(t + 1) is new. Here, we take $\tau = 0.1$.

Initialization (Ag(t)) includes N cells decided through the type of Ag(t). More precisely, if Ag(t) is new, such N cell are generated randomly; otherwise, some random B cells and $\eta\%$ of elements from the eventual population gotten in the last environment constitute the initial population. Here, we take $\eta = 10$.

Division (A, N_e) consists of two sub-populations S_1 and S_2, where each element in S_1 is superior to any one in S_2. Precisely, relying upon the concept of constraintdominance, population A is first sorted orderly into non-dominated classes F_1, F_2, ··· ,F_l, where the first is best and the last is worst. Suppose that F_1 involves in n_p members. If $n_p < N_e$, pick up some elements in F_2 with small concentrations, together with those in F_1, to constitute S_1, where the concentration of x in F_2, C(x), is decided by

$$C(x) = \frac{\left|\{y \in F_2 : \|x-y\| < \delta, y \prec_c x\}\right|}{\left|\{y \in A : y \prec_c x\}\right|} \tag{3}$$

Note that |A| denotes the number of elements in A, and we in this work take $\delta = 0.1n(t)$. If $|F_1 \cup F_2| < N_e$, go on this process by substituting F_3 for F_2 in Equation (3), until that S_1 includes N_e elements. If $n_p > N_e$, the conventional crowding distance approach, proposed by Deb et al. [2], is applied to F_1, and selects N_e elements with large crowding distances to form S_1. Further, S_2 consists of those elements in A but not in S_1.

Mutation (S_1, N) comprises of new B cells acquired through multiple mutation fashions on S_1. To this end, considering exploitation and exploration, we carry out a n_p-dependent mutation, where n_p is mentioned above. Namely, we first copy ξN better elements in S_1 to proliferate their clones with respectively clonal sizes m_c; each of clones undergoes the following Gaussian mutation with the mutation probability p_m,

$$x' \leftarrow x + \gamma N\left(0, (\xi\beta)^2\right) \tag{4}$$

with variance $\xi\beta$, where γ is a randomly generated number in (0, 1), and we take $\xi = 0.1$ in this paper. Further, for a pre-defined threshold λ with $0 < \lambda < 1$, if $n_p < \lambda N$, each B cell in S_1 changes its genes with the mutation probability $n(t)^{-1}$, relying upon Equation (4) but with variance α; otherwise, if a randomly generated number μ in (0, 1) satisfies $\mu < \lambda$, each element in S_1 shifts its genes as in the case of $n_p < \lambda N$ but with the variance, β and conversely, each B cell is mutated with the mutation probability p_m through

$$x' \leftarrow x + r^{r_1} \tag{5}$$

where r and r_1 are both randomly generated values in (0, 1). After the three mutation fashions are completed, all mutated cells constitute a new subpopulation B_1.

Recruitment (C) includes those better cells in C and some new ones obtained through crossover. That is, take η% of the worse elements in C to interact with B cells randomly selected in the above best subclass F_1, through crossover and with the crossover probability 1. After so, those new cells, together with the better cells in C, constitute the desired population A.

PERFORMANCE CRITERIA

To measure DCMMIOA's performance, several evaluation criteria are developed through extending the three conventional criteria [20], i.e., coverage rate (CR), coverage density (CD) and coverage scope (CS). If two algorithms A and B are executed respectively a single run in each of the T environments as in problem (VP) above, two series of non-dominated sets are acquired, $\{A^t\}_{t=1}^{T}$ and $\{B^t\}_{t=1}^{T}$.

(A) coverage rate. This can be used to compare the qualities of non-dominated solutions gotten by algorithms A and B, given by

$$CR(A,B) = \frac{1}{T}\sum_{t=1}^{T} \frac{\left|\left\{ y \in B^t \,\middle|\, \exists x \in A^t, s.t. x \prec_c y \right\}\right|}{\left|B^t\right|} \tag{6}$$

Equation (6) shows that the non-dominated sets found by algorithm A are globally better than those gained by algorithm B if $CR(A,B) > CR(B,A)$.

(B) coverage density and scope. Coverage density CD is utilized to measure the whole distribution of the non-dominated solutions obtained by algorithm A in the environments, defined as

$$CD = \frac{1}{T}\sum_{t=1}^{T}\sqrt{\frac{1}{|A^t|-1}\sum_{j=1}^{|A^t|}\left(\bar{d}^t - d_j^t\right)^2} \tag{7}$$

where

$$d_j^t = \min_{i \neq j, 1 \leq i \leq |A^t|} \left\{ \left\| \boldsymbol{f}(\boldsymbol{x}_i, t) - \boldsymbol{f}(\boldsymbol{x}_j, t) \right\| : \boldsymbol{x}_i, \boldsymbol{x}_j \in \boldsymbol{A}^t \right\}$$

$$\bar{d}^t = \frac{1}{|A^t|}\sum_{j=1}^{|A^t|} d_j^t$$

Equation (7) hints that the smaller the value CD, the better the distribution of the non-dominated solutions. Similarly, coverage scope is denoted by the average coverage width of solutions acquired in the environments, given by

$$CS = \frac{1}{T}\sum_{t=1}^{T} \max\left\{ \left\| \boldsymbol{f}(\boldsymbol{x}_i, t) - \boldsymbol{f}(\boldsymbol{x}_j, t) \right\| \middle| \boldsymbol{x}_i, \boldsymbol{x}_j \in A^t, 1 \leq i, j \leq |A^t| \right\} \tag{8}$$

Obviously, Equation (8) shows that a larger value of CS means a wider coverage scope of the solutions.

EXPERIMENTAL STUDY

In this section, we execute all experiments on a computer with CPU/3.0 GHz and RMB/2.0 GB. In order to test DCMMIOA's characteristics, three dynamic constrained multi-objective evolutionary algorithms reported are selected to compare against it, i.e., DMEA [6] and two similar versions of DNSGAII-A and DNSGAII-B [2]. In addition, we give seven DCMO problems DCTP1 to DCTP7 acquired by introducing a time-varying Rastrigin function into the seven static benchmark problems CTP1 to CTP7 [18], and two dynamic engineering design problems DSR and DPVM gained through modifying the coefficients of the two original versions [19,20]; more details can be found below. Such dynamic problems are utilized to examine the inherent properties of these four approaches.

To analyze fairly the performance characteristics of the above algorithms, each approach with population size 100 is admitted to execute 30 single runs on each test problem, and evaluates 10^5times in each environment for DCTP6, DSR and DPVM but 50,000 times for other test problems. Take four environments for each test problem. The other parameter settings of the three compared algorithms can be known through their corresponding references. For DCMMIOA, we take $m_c = 5$, $p_m = 0.3$, $N_e = 50$, $\alpha = 2.0$ for all the test problems; we set $\lambda = 0.8$, $\beta = 0.1$ for problems DCTP1 to DCTP5, DCTP7 and DSR, take $\lambda = 0.4$, $\beta = 2.0$ for DCTP6, and set $\lambda = 0.8$, $\beta = 2$ for DPVM.

Benchmark Problems and Experimental Analysis

Deb et al. designed a series of static constrained multiobjective test problems, i.e., CTP1 to CTP7 [18]. We modify their common function g(x) into the following time-varying Rastrigin's multimodal function,

$$g(x,t) = 1 + \sum_{i=2}^{n(t)} \left((x_i + 0.1t)^2 - 10\cos\left(2\pi(x_i + 0.1t)\right) + 10 \right),$$

$$x_1 \in [0,1], x_k \in [-5.12, 5.12], 2 \le k \le n(t)$$

and correspondingly, the static problems are extended in order into dynamic multi-objective multimodal problems DCTP1 to DCTP7. For either DCTP1 or DCTP7, the four environments are with dimensions 10, 10, 12 and 12 in turn, but 10, 10, 10 and 10 for each other problem. Depending on the performance criteria as in section 5, we obtain the statistical results in **Table 1** below, and the non-dominated fronts, found by such algorithms with respectively a single run, are presented in**Figure 1** below. In **Table 1**, A1, A2, A3 and A4 stand for DNSGAIIA, DNSGAII-B, DMEA and DCMMIOA, respectively.

Relying upon **Table 1** and **Figure 1**, DCMMIOA spends the least average runtime to solve all the test problems except DCTP4 and DCTP7, and DMEA is secondary. We also notice that DNSGAII-A and DNSGAII-B demand globally more time to deal with the above problems.

The average coverage rates (ACR) illustrate clearly that DCMMIOA's optimized quality for each of the above problems is globally best, as the qualities of the non-dominated solutions found by it in each environment are globally better than those gotten by any of the other algorithms for each problem, e.g., for DCTP1, the nondominated sets acquired by DCMMIOA in each environment cover averagely 38%, 40% and 90% of those obtained by DNSGAII-A, DNSGAII-B and DMEA, respectively; conversely, the non-dominated solutions acquired by the latter three algorithms in each environment dominate in order only 6%, 6% and 1% of those found by DCMMIOA. This shows that DCMMIOA, with a rational balance between convergence and diversity, can effectively track the global Pareto front in each environment per problem. In addition, **Figure 1** hints that once the environment becomes severe, DNSGAII-B and DMEA degrade seriously their performances; in particular, they can only acquire the local Pareto fronts (see Figures 1((a), (f) and (g)). Further, we note that only DCMMIOA can track the time-varying global Pareto front for DCTP6 (see **Figure 1**(f)), which shows that those compared algorithms get easily into local search and hence expose their weak diversity. The statistical values on CS present that each non-dominated set found by DCMMIOA in each environment for each problem has the globally wider coverage scope than that acquired by any of other algorithms, which indicates that such approach can find those excellent and diverse non-dominated solutions. In addition, the statistical values on CD show that these approaches can effectively eliminate those similar nondominated solutions, and thereby their resultant solutions are with the satisfactory distribution. Summarily, through the statistical values on ACR, CS and CD, we can clearly see that DCMMIOA acquires the best solution quality for each test problem while DNSGAII-A is secondary.

Table 1: Statistical comparison of non-dominated sets found for problems DCTP1 to DCTP7

Prob.		ACR(A, B) (%)				CD		CS		AT(s)
		A1	A2	A3	A4	*Mean*	*Std. Dev*	*Mean*	*Std. Dev*	
1	A1	0	30	85	6	0.05	0.03	0.98	10^{-2}	5.09
	A2	24	0	83	6	0.06	0.03	0.87	0.16	5.17
	A3	7	8	0	1	0.14	1.61	0.65	0.11	4.66
	A4	38	40	90	0	0.01	0.01	1.05	10^{-3}	3.64
2	A1	0	28	74	7	0.05	0.04	1.22	10^{-3}	4.57
	A2	24	0	72	5	0.06	0.05	1.21	10^{-4}	4.48
	A3	7	8	0	3	0.24	2.07	1.07	0.08	4.35
	A4	88	89	94	0	0.03	0.13	1.30	10^{-3}	3.66
3	A1	0	53	94	57	0.07	0.03	1.15	0.04	4.24
	A2	52	0	92	57	0.07	0.01	1.19	0.02	3.96
	A3	10	12	0	21	0.20	2.49	0.86	0.09	4.37
	A4	58	60	90	0	0.03	0.06	1.17	0.03	3.27
4	A1	0	47	87	35	0.08	0.02	0.86	0.07	3.23
	A2	58	0	93	44	0.07	0.01	0.82	0.04	3.13
	A3	12	8	0	11	0.20	1.32	0.48	0.11	3.95
	A4	75	69	92	0	0.01	10^{-3}	0.84	0.07	3.54
5	A1	0	37	85	25	0.07	0.02	1.18	10^{-2}	4.52
	A2	34	0	82	25	0.06	0.03	1.11	10^{-3}	4.33
	A3	7	10	0	11	0.22	1.27	0.88	0.08	4.25
	A4	78	79	87	0	0.02	0.01	1.12	0.01	3.54
6	A1	0	26	48	20	0.06	0.07	1.15	0.91	10.1
	A2	29	0	50	15	0.08	0.09	1.24	0.81	9.93
	A3	7	8	0	6	0.11	0.28	0.85	1.13	7.17
	A4	43	38	58	0	0.02	10^{-5}	1.77	0.01	6.33
7	A1	0	42	81	2	0.15	0.30	0.85	0.21	4.01
	A2	43	0	80	2	0.13	0.08	0.78	0.31	4.11
	A3	18	17	0	1	0.18	0.47	0.33	0.36	3.36
	A4	80	79	99	0	0.11	0.05	1.63	0.07	4.53

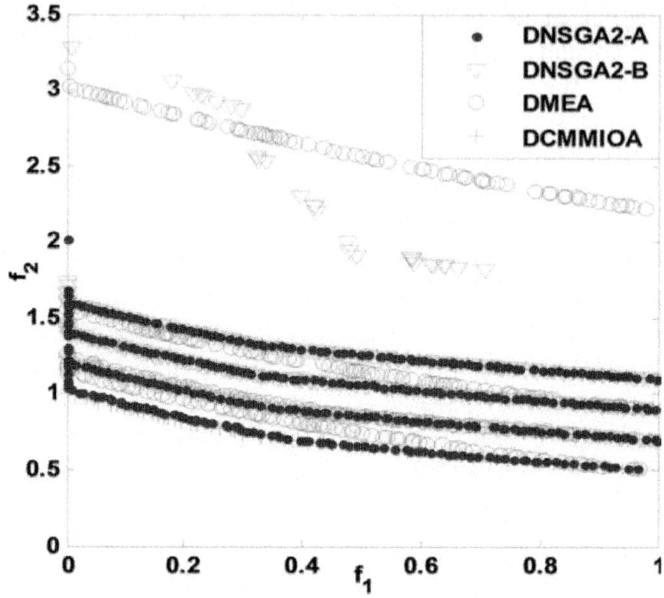
3.5
3
2.5
2
1.5
1
0.5
0
f2
0
0.2
0.4
f1
0.6
0.8
1
DNSGA2-A
DNSGA2-B
DMEA
DCMMIOA

(a)

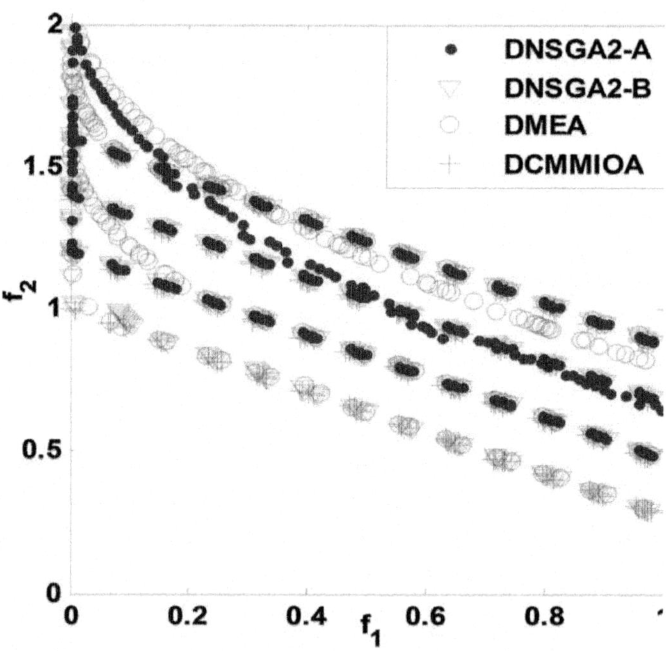
2
1.5
1
0.5
0
f2
0
0.2
0.4
f1
0.6
0.8
DNSGA2-A
DNSGA2-B
DMEA
DCMMIOA

(b)

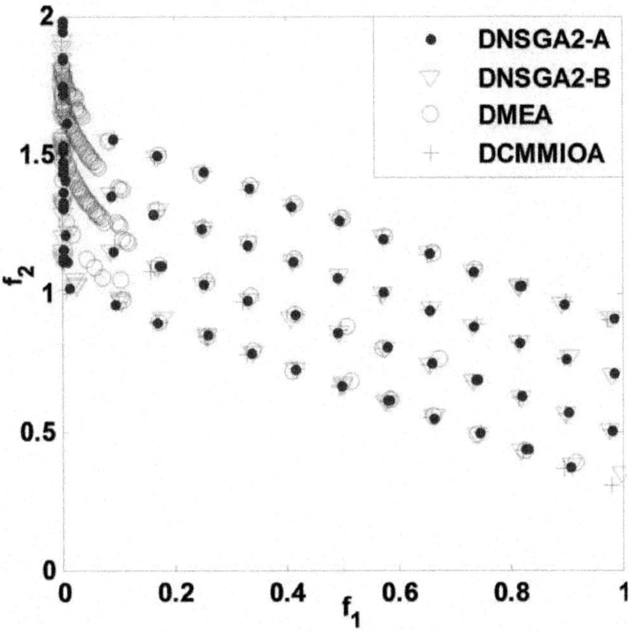
DNSGA2-A
DNSGA2-B
DMEA
DCMMIOA
f2
f1
0
0.5
1
1.5
2
0.2
0.4
0.6
0.8

(c)

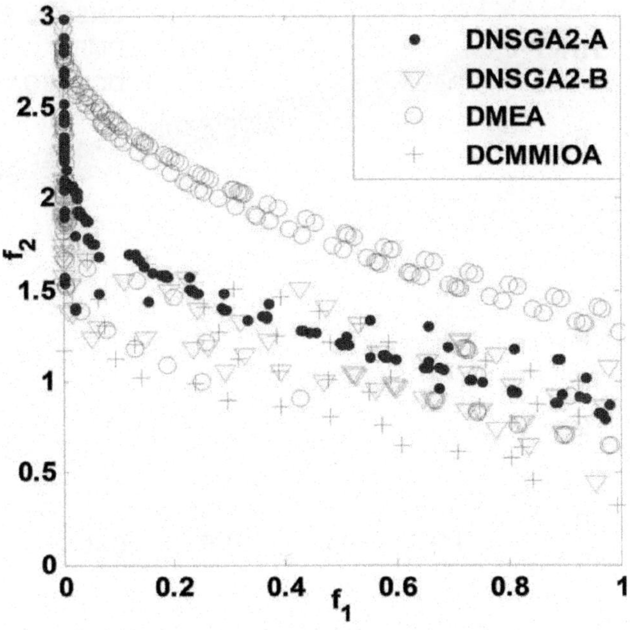
DNSGA2-A
DNSGA2-B
DMEA
DCMMIOA
f2
f1
0
0.5
1
1.5
2
2.5
3
0.2
0.4
0.6
0.8

(d)

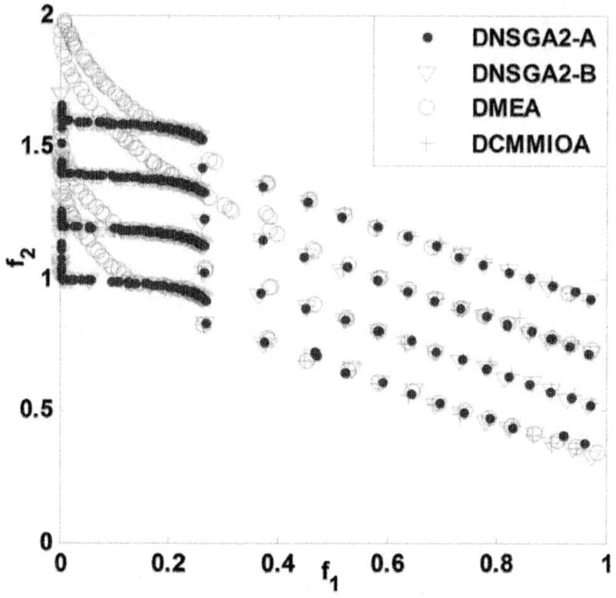
DNSGA2-A
DNSGA2-B
DMEA
DCMMIOA
f2
f1
0
0.5
1
1.5
2
0.2
0.4
0.6
0.8

(e)

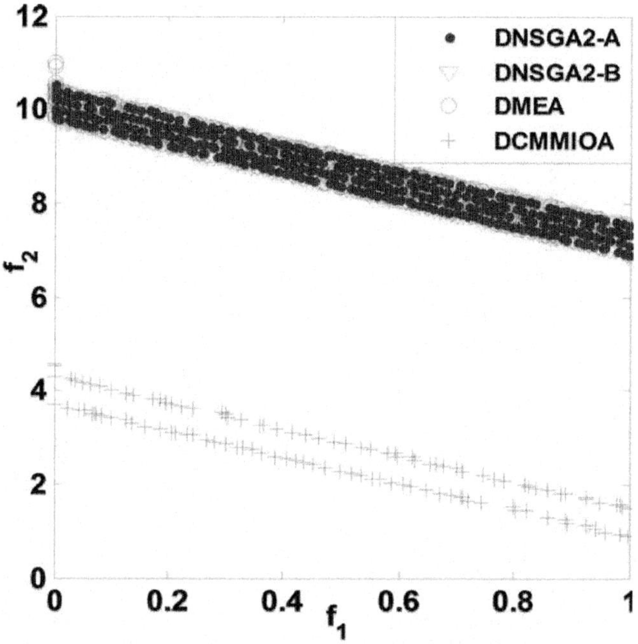
DNSGA2-A
DNSGA2-B
DMEA
DCMMIOA
f2
f1
0
2
4
6
8
10
12
0.2
0.4
0.6
0.8
1

(f)

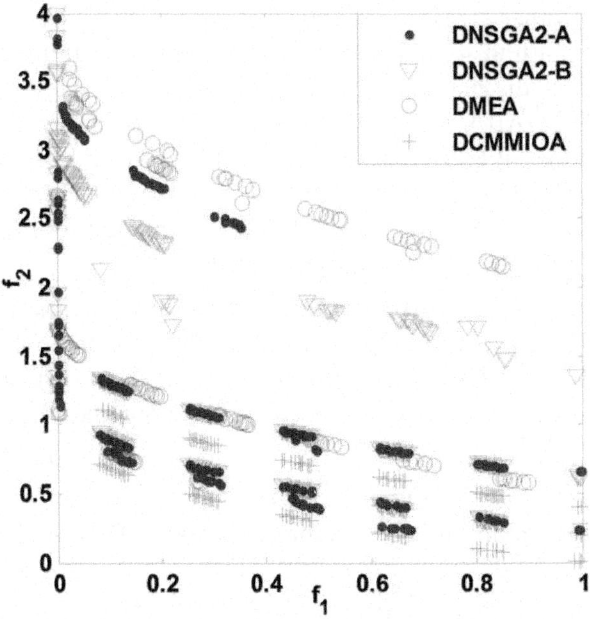

(g)

Figure 1: Comparison of non-dominated fronts obtained by the four algorithms in each environment. (a) DCTP1; (b) DCTP2; (c) DCTP3; (d) DCTP4; (e) DCTP5; (f) DCTP6; (g) DCTP7.

Totally, for each test problem above, we can draw the strong conclusion that DCMMIOA performs well over the compared approaches; DMEA has the higher efficiency than either DNSGAII-A or DNSGAII-B, but the worse effect than DNSGAII-A.

Experimental Results on Engineering Problems

We take the dynamic speed reducer model (DSR) for example. Such model can be found in [19]. Similar to the above experiment, the above algorithms get their statistical results given in **Table 2**, and the non-dominated fronts are drawn in **Figure 2**. Note that since DMEA fails to solve such model, we do not give its statistical results.

Table 2 and **Figure 2** display the results obtained by DNSGAII-A, DNSGAII-B and DCMMIOA for problem DSR. We know easily that DCMMIOA has the prominent superiority over two other algorithms, because the non-dominated sets obtained by it cover 96% of those found by either DNSGAII-A or DNSGAII-B, i.e., ACR (A4, A1) = 96%, ACR(A4, A2) = 96%.

Further, from other statistical values, we observe that in each environment, DCMMIOA can find some non-dominated solutions with the wide coverage and satisfactory distribution. On the other hand, we see that the solutions found by DNSGAIIA and DNSGAII-B have the similar distribution and coverage. Further, it is obvious that DCMMIOA has the high efficiency when solving the above problem, but the compared approaches cause the slow searching behaviors. Totally, DCMMIOA performs best for this problem, and DNSGAII-B is slightly better than DNSGAII-A.

We next solve the other engineering problem, Dynamic Pressure Vessel Model (DPVM). The static pressure vessel model is proposed originally by Deb and Srinivasan [20]. It aims to minimize the cost of fabrication and to maximize the storage capacity of the vessel, and includes four variables, i.e., thickness of cylindrical part of the vessel (x_1), the hemispherical heads (x_2), radius of vessel (x_3) and length of vessel (x_4). We convert it into a dynamic model through modifying the parameter settings. DPVM is formulated as follow:

$$\min_{x \in R^n} \boldsymbol{f}(\boldsymbol{x},t) = \left(f_1(\boldsymbol{x},t), f_2(\boldsymbol{x},t)\right)$$

$$s.t., f_1(\boldsymbol{x},t) = a_1(t)x_1x_4x_3 + a_2(t)x_2x_3^2 + a_3(t)x_1^2x_4^2$$

$$+a_4(t)x_1^2x_3^2, f_2(\boldsymbol{x},t) = -\left(\pi x_3^2 x_4 + a_5(t)\pi x_3^3\right),$$

$$g_1 : 0.0193x_3 - x_1 \le 0, g_2 : 0.00954x_3 - x_2 \le 0,$$

$$0.0625 \le x_1, x_2 \le 5, 10 \le x_3 \le 200, 10 \le x_4 \le 240$$

where the coefficients $a_i(t)$, $1 \le i \le 5$, are acquired by introducing time-varying parameters,

$$a_i(t) = a_i + \frac{t}{10}, 1 \le i \le 5, a_1 = 0.6224, a_2 = 1.7781,$$

$$a_3 = 1.58305, a_4 = 9.92, a_5 = 1.333.$$

Similar to the process of solving the above problems, the algorithms obtain their statistical results for DPVM given in **Table 3**, and their non-dominated fronts are presented in **Figure 3** below.

Through **Table 3** and **Figure 3**, we notice that although DMEA behaves worst, it can obtain some feasible solutions in each environment. Besides, we can obtain the same conclusion as that given in DSR; namely, by comparison to either DNSGAII-A or DNSGAII-B, DCMMIOA has the best optimized quality and the higher performance efficiency; DNSGAII-A and DNSGAII-B present similar effects and almost equal efficiencies. On the other hand, we also see that the non-dominated sets found by DNSGAII-A, DNSGAII-B and DCMMIOA are with similar distributions and coverage scopes, but there are some subtle differences between them. DCMMIOA can globally achieve stable search performance, due to the small variances on CD and CS. Therefore, for the above problem, DCMMIOA performs best, and DNSGAIIA and DNSGAII-B are secondary.

Table 2: Comparison of statistical values obtained for DSR.

	ACR(A, B) (%)			CD		CS		AT(s)
	A1	A2	A3	Mean	*Std. Dev*	Mean	*Std. Dev*	
A1	0	14	0.4	13.99	1.029	1408.4	2.116	12.67
A2	15	0	0.6	14.00	0.486	1408.5	1.279	12.63
A4	96	96	0	12.88	1.13	1438.5	1.505	8.51

Table 3: Comparison of statistical values obtained by the four algorithms for problem DPVM

	ACR(A, B) (%)				CD		CS		AT(s)
	A1	A2	A3	A4	Mean	Var	Mean	Var	
A1	0	20	53	19	563447	8.88×10^8	6.78×10^7	6.88×10^9	11.98
A2	20	0	51	19	558483	5.51×10^8	6.72×10^7	1.56×10^{10}	11.86
A3	4	4	0	6	743289	7.17×10^{10}	5.50×10^7	3.26×10^{14}	7.21
A4	43	43	55	0	578021	6.3×10^9	6.69×10^7	7.65×10^9	8.24

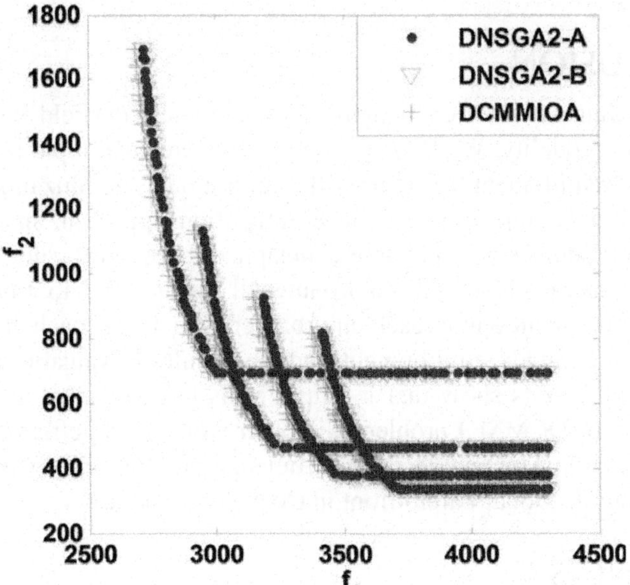

Figure 2: Comparison of non-dominated fronts obtained in each environment for DSR.

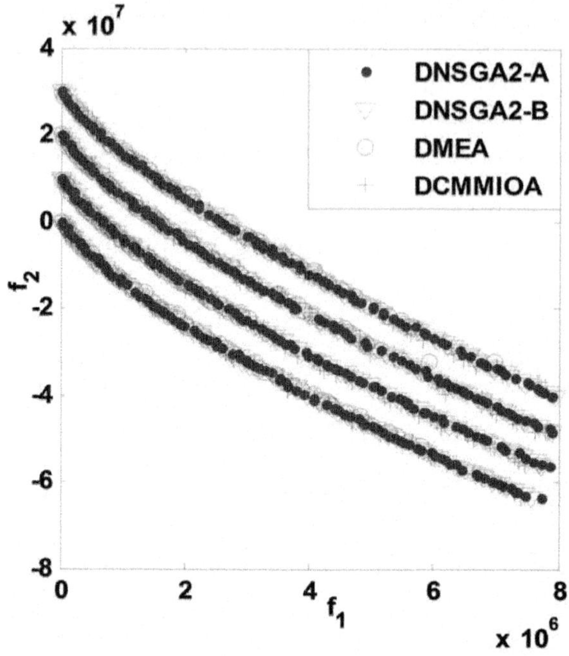

Figure 3: Comparison of non-dominated fronts obtained in each environment for DPVM.

CONCLUSION

DCMMO is an extremely challenging research topic in the field of optimization, due to multimodality. We in this paper investigate such topic because of the increasing requirement of real-world multimodal optimization problems. Correspondingly, one bio-immune optimization approach, inspired by the immune surveillance and interactive metaphors between B cells, is developed to adapt dynamically to the environmental change and to find the global Pareto optimal solutions in each environment. Such approach can adaptively monitor the complex environment and find multiple valuable regions. The experimental results show that it outperforms the compared algorithms and is potential for DCMMO problems; in other word, it can efficiently track the dynamic environment and keep the sufficient diversity, while being capable of approaching the global Pareto front in each environment.

REFERENCES

1. M. Farina, K. Deb and P. Amato, "Dynamic Multi-Objective Optimization Problems: Test Case, Approximations, and Applications,"

Evolutionary Computation, Vol. 8, No. 5, 2004, pp. 425-442. doi:10.1109/TEVC.2004.831456

2. K. Deb, B. R. N. Udaya and S. Karthik, "Dynamic MultiObjective Optimization and Decision-Making Using Modified NSGA-II: A Case Study on Hydro-Thermal Power Scheduling Bi-Objective Optimization Problems," Lecture Notes in Computer Science, Vol. 4403, 2007, pp. 803-817. doi:10.1007/978-3-540-70928-2_60
3. J. Mehnen, T. Wagner and G. Rudolph, "Evolutionary Optimization of Dynamic Multi-Objective Test Functions," Proceedings of the Second Italian Workshop on Evolutionary Computation, Siena, September 2006.
4. A. Zhou, Y. C. Jin, Q. Zhang, B. Sendhoff and E. Tsang, "Prediction-Based Population Re-Initialization for Evolutionary Dynamic Multi-Objective Optimization," The Fourth International Conference on Evolutionary MultiCriterion Optimization, Matsushima, 5-8 March 2007, pp. 832-846.
5. I. Hatzakis and D. Wallace, "Dynamic Multi-Objective Optimization with Evolutionary Algorithms: A ForwardLooking Approach," Proceedings of the 8th Annual Conference on Genetic and Evolutionary Computation, Seattle, Washington, 2006, pp. 1201-1208.
6. C. A. Liu and Y. P. Wang, "Multi-Objective Evolutionary Algorithm for Dynamic Nonlinear Constrained Optimization Problems," Systems Engineering and Electronics, Vol. 20, No. 1, 2009, pp. 204-210.
7. K. C. Tan, "A Competitive-Cooperative Coevolutionary Paradigm for Dynamic Multi-Objective Optimization," IEEE Transactions on Evolutionary Computation, Vol. 13, No. 1, 2009, pp. 103-127. doi:10.1109/TEVC.2008.920671
8. R. H. Shang, L. C. Jiao, M. G. Gong and B. Lu, "Clonal Selection Algorithm for Dynamic Multi-Objective Optimization," In: Y. Hao, et al., Eds., Computational Intelligence and Security, Springer, Berlin, Heidelberg, Vol. 3801, 2005, pp. 846-851.doi:10.1007/11596448_125
9. Z. H. Zhang, "Multi-Objective Optimization Immune Algorithm in Dynamic Environments and Its Application to Greenhouse Control," Applied Soft Computing, Vol. 8, No. 2, 2008, pp. 959-971. doi:10.1016/j.asoc.2007.07.005
10. Z. H. Zhang and S. Q. Qian, "Artificial Immune System in Dynamic Environments Solving Time-Varying NonLinear Constrained Multi-Objective Problems," Soft Computing, Vol. 15, No. 7, 2011, pp. 1333-1349. doi:10.1007/s00500-010-0674-z
11. E. Hart and J. Timmis, "Application Areas of AIS: The Past, Present and the Future," Applied Soft Computing, Vol. 8, No. 1, 2008, pp. 191-201.

doi:10.1016/j.asoc.2006.12.004

12. J. Timmis, P. Andrews and E. Hart, "On Artificial Immune Systems and Swarm Intelligence," Swarm Intelligence, Vol. 4, No. 4, 2010, pp. 247-273. doi:10.1007/s11721-010-0045-5
13. C. A. Coello Coello and N. Cruz Cortés, "An Approach to Solve Multi-Objective Optimization Problems Based on an Artificial Immune System," In: J. Timmis and P. J. Bentley, Eds., First International Conference on Artificial Immune Systems, Canterbury, 2002, pp. 212-221.
14. N. Cruz Cortés and C. A. Coello Coello, "Using Artificial Immune Systems to Solve Optimization Problems," Workshop Program of Genetic and Evolutionary Computation Conference, 2003, pp. 312-315.
15. I. Aydin, M. Karakose and E. Akin, "A Multi-Objective Artificial Immune Algorithm for Parameter Optimization in Support Vector Machine," Applied Soft Computing, Vol. 11, No. 1, 2011, pp. 120-129. doi:10.1016/j.asoc.2009.11.003
16. Z. H. Hu, "A Multiobjective Immune Algorithm Based on a Multiple-Affinity Model," European Journal of Operational Research, Vol. 202, No. 1, 2010, pp. 60-72.doi:10.1016/j.ejor.2009.05.016
17. J. Q. Gao and J. Wang, "WBMOAIS: A Novel Artificial Immune System for Multi-Objective Optimization," Computers & Operations Research, Vol. 37, No. 1, 2010, pp. 50-61.doi:10.1016/j.cor.2009.03.009
18. K. Deb, A. Pratap, and T. Meyarivan, "Constrained Test Problems for Multi-Objective Evolutionary Optimization," Evolutionary Multi-Criterion Optimization Lecture Notes in Computer Science, Vol. 1993, 2001, pp. 284- 298.
19. A. Kurpati, S. Azarm and J. Wu, "Constraint Handling Improvements for Multiobjective Genetic Algorithms," Structural Multidisciplinary Optimization, Vol. 23, No. 3, 2002, pp. 204-213. doi:10.1007/s00158-002-0178-2
20. K. Deb and A. Srinivasan, "Monotonicity Analysis, Evolutionary Multi-Objective Optimization, and Discovery of Design Principles," Indian Institute of Technology, Kanpur Genetic Algorithm Laboratory KanGAL), Report No. 2006004, 2006.

Chapter 2

A NEW APPROACH TO THE OPTIMIZATION OF THE CVRP THROUGH GENETIC ALGORITHMS

Mariano Frutos[1], Fernando Tohmé[2]

[1]Department of Engineering, National University of the South, Bahía Blanca, Argentina

[2]Department of Economics, National University of the South, Bahía Blanca, Argentina

ABSTRACT

This paper presents a new approach to the analysis of complex distribution problems under capacity constraints. These problems are known in the literature as CVRPs (Capacitated Vehicle Routing Problems). The procedure introduced in this paper optimizes a transformed variant of a CVRP. It starts generating feasible clusters and codifies their ordering. In the next stage the procedure feeds this information into a genetic algorithm for its optimization. This makes the algorithm independent of the constraints and improves its performance. Van Breedam problems have been used to test this technique. While the results obtained are similar to those in other works, the processing times are longer.

INTRODUCTION

The complexity of distribution problems is the reason for the lack of solution techniques able to provide optimal solutions to real cases in a reasonable time [1]. But firms have to make decisions on transportation matters, sometimes in the short run. The economic importance of moving people and goods makes it necessary the development of strategies to generate quick competitive solutions [2,3]. In the literature, these problems are generically known as the Vehicle Routing Problem (VRP) [4], which is NP-Complete combinatorial optimization problem. Its scholarly and practical interest stems from the aforementioned lack of quick solution methods [5]. The main goal of this paper is to introduce a new algorithm, robust, flexible and fast enough to solve different instances of this problem.

The rest of the paper is organized as follows. Firstly, we present some characterizations of the VRP as well as of the original version of the CVRP (Capacitated Vehicle Routing Problem). Then, we introduce an alternative version of the latter and a methodology for its solution. We present later the result of the experiments with this method. Finally we discuss possible further work on the subject.

DEFINITION OF CVRP

In rough terms, the statement of the VRP is: given classes of clients and of storage sites (both scattered over a geographical region) and a fleet of vehicles, determine the minimal cost routes starting and ending in storage sites delivering the goods to the clients. The features of clients, storage sites and vehicles, as well as the operational constraints over routes determine the different variants of the problem. Each client is associated to a demand that has to be covered by a vehicle. This demand is usually understood as an amount of a good that occupies some volume of the vehicle. In crowded area a single vehicle may not be able to satisfy the demands of all the clients living there. The capacity of the vehicle imposes then a bound on the number of clients that it might satisfy at the same time [6,7]. This kind of constraints defines the Capacitated Vehicle Routing Problem (CVRP). The capacity of the vehicle might be in terms of weight or volume. When more than a single kind of products is transported the vehicles could be compartmented carrying each type of parcel in a different subdivision. A distinction can be made between the problems of homogeneous fleets, in which the attributes (capacity, cost, etc.) are the same for all the vehicles, and heterogeneous fleet problems. The latter case raises the possibility of incompatibilities between the capacities of the vehicles and the demands of the clients. If then, each client might be visited only by some of the vehicles. It could also happen that the goods to be transported might not be available in a single storage site and must be collected from several sites. All these depots have to be visited before delivering the goods to the client. While normally each client should be visited once, it could be necessary to satisfy his demand at different moments and by different vehicles. Some clients could have some hourly restrictions for the reception of parcels, which are expressed as time windows [8,9]. In such case the problem is known as Vehicle Routing Problem with Time Windows (VRPTW). The vehicles, as well as the products, are usually parked at the storage sites. It is also usually required that the vehicle starts and ends trips from the same storage site, but this restriction can be lifted in some cases. The number of available vehicles could be given or be decision variable [10]. The objective is frequently the minimization of the distances as well as the number of vehicles used while satisfying the demands of the clients

[11,12]. Recent studies have allowed the possibility of vehicles traversing different routes [13]. Different presentations of such problems can be found in the literature [14-17], as well as different approaches to solving them [18-23].

In this work we focus on the CVRP. Consider version (1) of the problem. Let N^c be the class of client nodes, endowed with a linear order <, N^s the set of storage nodes while $N = N^c \cup N^s$. C_{ij} the cost (distance) of going from nodes i to j for any pair $(i,j) \in N \times N$. X_{ij} is a binary variable, which is 1 if a vehicle goes from i to j without intermediate stops. Otherwise $X_{ij} = 0$. $C_{\max}$ is the maximum capacity of a vehicle and d_i the amount demanded at a node $i \in N^c$. Let S_j be the class of nodes in a path that starts and ends at node $j \in N^s$. Then, the problem is:

$$\min \sum_{i \in N} \sum_{j \in N} C_{ij} X_{ij} \tag{1}$$

s.t.:

1) $\sum_{i \in N^c \cap S_j} d_i \leq C_{\max}$ 2) and $\{S_j\}_{j \in N^s}$ such that $(S_i \cap S_j) \cap N^c = \varnothing$

In words: minimize the total distance traversed by all vehicles, satisfying the demands on every route for a partition of client nodes in clusters, each one covered by a single vehicle. Notice that constraint 2) makes the problem more complex than a linear programming one.

Version (2) is as follows: let $M = \{S_j\}_{j \in N^s}$ for some allocation of routes, $c_m = \sum_{(i,j) \in m} C_{ij}$ the sum of distances over route m, X_m is a binary variable that is 1 if route m is used and 0 otherwise and e_{im} also a binary variable (1 if $i \in N^c$ belongs to route m and 0 otherwise). Then, the problem is:

$$\min \sum_{m \in M} c_m X_m \tag{2}$$

s.t.:

3) and for each i, $\sum_{m \in M} e_{im} X_m = 1$ That is (2) is analogous to (1) under constraint 1) with constraint 2) already satisfied.

METHODOLOGY

An algorithm solving (2) proceeds in two stages. The first one generates feasible clusters, classifying and ordering them. The second stage solves the minimization problem by means of a Genetic Algorithm [24].

First Stage

The algorithm generates the clusters that will be elements of $\{S_j\}_{j \in N^s}$. We take each $i \in N^c$ and consider all the clusters S_j such that $i \in S_j$. The inductive construction starts with i = 1 (the lowest in the < order) building all the clusters of client nodes that include 1, satisfying condition 1). Then, if all possible clusters up to i = k have been obtained, consider i = k +1 (where, of course, k < k + 1). All routes starting and ending in a node of N^s that go through client nodes in $N^c - \{1,\cdots,k\}$, including node k + 1, satisfying condition 1), are constructed.

For each $i \in N^c$ its class of clusters, denoted S^i, is endowed with a weak order $Đ_i$ (reflexive, antisymmetric and transitive, i.e. allowing ties) according to the number of clients in each cluster. Among clusters covering the same number of client nodes, say $\{1,\cdots,i,\cdots,k\}$ and $\{1,\cdots,i,\cdots,l\}$, if k < l then $\{1,\cdots,i,\cdots,k\}$ $Đ_i$ $\{1,\cdots,i,\cdots,l\}$. The highest rank in $Đ_i$ corresponds to routes covering i as well as the largest possible number of other client nodes in the upper level of order <. As an example, consider the case of 10 client nodes demanding 50 units and a vehicle capacity of 100 units. **Table 1** shows the distances among nodes, being 0 the storage site from which departs the vehicle. **Table 2** shows the result of the first stage of the procedure. Notice that, since the capacity of the vehicle is 100 and each client node demands 50, a route that clusters some clients together can only consist of two of them. So, for instance the group corresponding to node 9 includes only two possible clusters, {9} and {9,10}, because all other nodes have already been assigned to other clusters. Cluster {9,10} is of a higher order than {9} because it includes two clients instead of just one node.

Second Stage

At this point a Genetic Algorithm [24] is applied, designed for its use in this problem. An individual is codified in a chromosome constructed in such way as to satisfy conditions 2) and 3) and therefore are feasible under problem (2). The coding operates on a lexicographic basis. Each $i \in N^c$ is given a number, starting with i = 1, and following the process according to order <. So, take all the clusters ordered according to $Đ_1$. This order can be represented by the linear numerical order $1,\cdots,|S^1|$. So, if the k-th cluster in the ordering is chosen, the corresponding gene in the chromosome is k. For i = j, the order $Đ_j$ is applied to a subset of S^j obtained by eliminating ("filtered out") all the clusters that include nodes already present in clusters corresponding to genes i = 1,···, j – 1. The new ordering is denoted $\angle i'$.

Table 1: Distances—Case: 10 nodes / Demand per node: 50 units / Vehicle capacity: 100 units.

Nodes	0	1	2	3	4	5	6	7	8	9	10
0	0.00	5.00	14.14	7.07	11.18	15.81	14.14	5.00	11.18	11.18	18.03
1	5.00	0.00	11.18	11.18	15.81	20.62	18.03	10.00	7.07	10.00	22.36
2	14.14	11.18	0.00	15.81	25.00	25.50	20.00	18.03	5.00	20.62	32.02
3	7.07	11.18	15.81	0.00	11.18	10.00	7.07	5.00	15.00	18.03	18.03
4	11.18	15.81	25.00	11.18	0.00	11.18	15.00	7.07	22.36	15.81	7.07
5	15.81	20.62	25.50	10.00	11.18	0.00	7.07	11.18	25.00	25.00	15.00
6	14.14	18.03	20.00	7.07	15.00	7.07	0.00	11.18	20.62	25.00	20.62
7	5.00	10.00	18.03	5.00	7.07	11.18	11.18	0.00	15.81	14.14	14.14
8	11.18	7.07	5.00	15.00	22.36	25.00	20.62	15.81	0.00	15.81	29.15
9	11.18	10.00	20.62	18.03	15.81	25.00	25.00	14.14	15.81	0.00	20.00
10	18.03	22.36	32.02	18.03	7.07	15.00	20.62	14.14	29.15	20.00	0.00

Table 2: Clusters—Case: 10 nodes / Demand per node: 50 units / Vehicle capacity: 100 units

Nodes	Clusters									
0	-	-	-	-	-	-	-	-	-	-
1	1	1.2	1.3	1.4	1.5	1.6	1.7	1.8	1.9	1.10
2	2	2.3	2.4	2.5	2.6	2.7	2.8	2.9	2.10	-
3	3	3.4	3.5	3.6	3.7	3.8	3.9	3.10	-	-
4	4	4.5	4.6	4.7	4.8	4.9	4.10	-	-	-
5	5	5.6	5.7	5.8	5.9	5.10	-	-	-	-
6	6	6.7	6.8	6.9	6.10	-	-	-	-	-
7	7	7.8	7.9	7.10	-	-	-	-	-	-
8	8	8.9	8.10	-	-	-	-	-	-	-
9	9	9.10	-	-	-	-	-	-	-	-
10	10	-	-	-	-	-	-	-	-	-

Then the j-th gene in the chromosome is the rank in $\angle i'$ of the cluster that is chosen. If for a given i the domain obtained is Æ (because all possible nodes belong to clusters corresponding to genes 1 to i – 1), any number between 1 and $|S^{i}|$ can be assigned to the i-th gene. In this way, a chromosome is defined, each corresponding to a feasible solution to problem (2). The Genetic Algorithm seeks to find the optimal one.

We can see some examples of how chromosomes encode clusters in **Table 3**, (3-3-2-1-4-2-3-2-1-1) and **Table 4** (5-2-1-5-1-4-2-2-1-1): all routes start and end at node 0 and go through two client nodes; in gray we see the filtered out clusters and in red the selection. Consider for instance the second gene in **Table 3**: it has to discard (written in gray) all routes that go through node 3, since it has already been assigned to the first gene. Number 3 assigned to this gene indicate that the route selected is the third in the list of non-filtered routes. On the other hand, a number assigned to a gene with a row with all routes filtered out does not have any effect. The initial population consists of individuals whose chromosomes are constituted by genes selected at random.

Once established a chromosome, it is decoded and evaluated. This evaluation is performed according to problem (2). For the computation of each cluster we use Lin-Kerninghan's algorithm that yields the minimal distance among its nodes. The following operators have been chosen in the light of some previous evaluations. The cross-over operator is the two-point one (**Table 5**, Parent 1: 3-3- 2-_-_-_-_-2-1-1, Parent 2: _-_-_-5-1-4-2-_-_-_, Offspring: 3-3-2-5-1-4-2-2-1-1) and the mutation operator alters gene values at 10% of the chromosome (Mutation (**Table 6**, Offspring*: 3-3-2-5-1-2-2-2-1-1)). Finally, the selection is by ranking.

The Genetic Algorithm runs until the solution gets stabilized. These tests were implemented with a C^{++} program running on a 3.00 GHZ CPU with a 2.00 GB RAM. The parameters of the experiments were as follows. The size of the population: 100; the cross-over probability: 0.70 and the probability of mutation: 0.01.

Before running the algorithm over problems in the literature we experimented over this example.**Figure 1** shows the nodes and their proportional distances in a two-dimensional framework. Figures 2 to 7 show the routes generated in several generations and the minimal total distance covered by the selected routes.

Table 3: Example of a chromosome (3-3-2-1-4-2-3-2-1-1) that will act as parent 1 in the cross-over of **Table 5**.

Chr.	Nodes	Clusters									
3	1	0-1-0	0-1-2-0	0-1-3-0	0-1-4-0	0-1-5-0	0-1-6-0	0-1-7-0	0-1-8-0	0-1-9-0	0-1-10-0
3	2	0-2-0	0-2-3-0	0-2-4-0	0-2-5-0	0-2-6-0	0-2-7-0	0-2-8-0	0-2-9-0	0-2-10-0	-
2	3	0-3-0	0-3-4-0	0-3-5-0	0-3-6-0	0-3-7-0	0-3-8-0	0-3-9-0	0-3-10-0	-	-
1	4	0-4-0	0-4-5-0	0-4-6-0	0-4-7-0	0-4-8-0	0-4-9-0	0-4-10-0	-	-	-
4	5	0-5-0	0-5-6-0	0-5-7-0	0-5-8-0	0-5-9-0	0-5-10-0	-	-	-	-
2	6	0-6-0	0-6-7-0	0-6-8-0	0-6-9-0	0-6-10-0	-	-	-	-	-
3	7	0-7-0	0-7-8-0	0-7-9-0	0-7-10-0	-	-	-	-	-	-
2	8	0-8-0	0-8-9-0	0-8-10-0	-	-	-	-	-	-	-
1	9	0-9-0	0-9-10-0	-	-	-	-	-	-	-	-
1	10	0-10-0	-	-	-	-	-	-	-	-	-

Table 4: Example of a chromosome (5-2-1-5-1-4-2-2-1-1) that will act as parent 2 in the cross-over of **Table 5**.

Chr.	Nodes	Clusters									
5	1	0-1-0	0-1-2-0	0-1-3-0	0-1-4-0	0-1-5-0	0-1-6-0	0-1-7-0	0-1-8-0	0-1-9-0	0-1-10-0
2	2	0-2-0	0-2-3-0	0-2-4-0	0-2-5-0	0-2-6-0	0-2-7-0	0-2-8-0	0-2-9-0	0-2-10-0	-
1	3	0-3-0	0-3-4-0	0-3-5-0	0-3-6-0	0-3-7-0	0-3-8-0	0-3-9-0	0-3-10-0	-	-
5	4	0-4-0	0-4-5-0	0-4-6-0	0-4-7-0	0-4-8-0	0-4-9-0	0-4-10-0	-	-	-
1	5	0-5-0	0-5-6-0	0-5-7-0	0-5-8-0	0-5-9-0	0-5-10-0	-	-	-	-
4	6	0-6-0	0-6-7-0	0-6-8-0	0-6-9-0	0-6-10-0	-	-	-	-	-
2	7	0-7-0	0-7-8-0	0-7-9-0	0-7-10-0	-	-	-	-	-	-
2	8	0-8-0	0-8-9-0	0-8-10-0	-	-	-	-	-	-	-
1	9	0-9-0	0-9-10-0	-	-	-	-	-	-	-	-
1	10	0-10-0	-	-	-	-	-	-	-	-	-

Table 5: Cross-over (offspring: 3-3-2-5-1-4-2-2-1-1).

Chr.	Nodes	Clusters									
3	1	0-1-0	0-1-2-0	0-1-3-0	0-1-4-0	0-1-5-0	0-1-6-0	0-1-7-0	0-1-8-0	0-1-9-0	0-1-10-0
3	2	0-2-0	0-2-3-0	0-2-4-0	0-2-5-0	0-2-6-0	0-2-7-0	0-2-8-0	0-2-9-0	0-2-10-0	-
2	3	0-3-0	0-3-4-0	0-3-5-0	0-3-6-0	0-3-7-0	0-3-8-0	0-3-9-0	0-3-10-0	-	-
5	4	0-4-0	0-4-5-0	0-4-6-0	0-4-7-0	0-4-8-0	0-4-9-0	0-4-10-0	-	-	-
1	5	0-5-0	0-5-6-0	0-5-7-0	0-5-8-0	0-5-9-0	0-5-10-0	-	-	-	-
4	6	0-6-0	0-6-7-0	0-6-8-0	0-6-9-0	0-6-10-0	-	-	-	-	-
2	7	0-7-0	0-7-8-0	0-7-9-0	0-7-10-0	-	-	-	-	-	-
2	8	0-8-0	0-8-9-0	0-8-10-0	-	-	-	-	-	-	-
1	9	0-9-0	0-9-10-0	-	-	-	-	-	-	-	-
1	10	0-10-0	-	-	-	-	-	-	-	-	-

Table 6: Mutation (offspring*: 3-3-2-5-1-4-2-2-1-1).

Chr.	Nodes	Clusters									
3	1	0-1-0	0-1-2-0	0-1-3-0	0-1-4-0	0-1-5-0	0-1-6-0	0-1-7-0	0-1-8-0	0-1-9-0	0-1-10-0
3	2	0-2-0	0-2-3-0	0-2-4-0	0-2-5-0	0-2-6-0	0-2-7-0	0-2-8-0	0-2-9-0	0-2-10-0	-
2	3	0-3-0	0-3-4-0	0-3-5-0	0-3-6-0	0-3-7-0	0-3-8-0	0-3-9-0	0-3-10-0	-	-
5	4	0-4-0	0-4-5-0	0-4-6-0	0-4-7-0	0-4-8-0	0-4-9-0	0-4-10-0	-	-	-
1	5	0-5-0	0-5-6-0	0-5-7-0	0-5-8-0	0-5-9-0	0-5-10-0	-	-	-	-
2	6	0-6-0	0-6-7-0	0-6-8-0	0-6-9-0	0-6-10-0	-	-	-	-	-
2	7	0-7-0	0-7-8-0	0-7-9-0	0-7-10-0	-	-	-	-	-	-
2	8	0-8-0	0-8-9-0	0-8-10-0	-	-	-	-	-	-	-
1	9	0-9-0	0-9-10-0	-	-	-	-	-	-	-	-
1	10	0-10-0	-	-	-	-	-	-	-	-	-

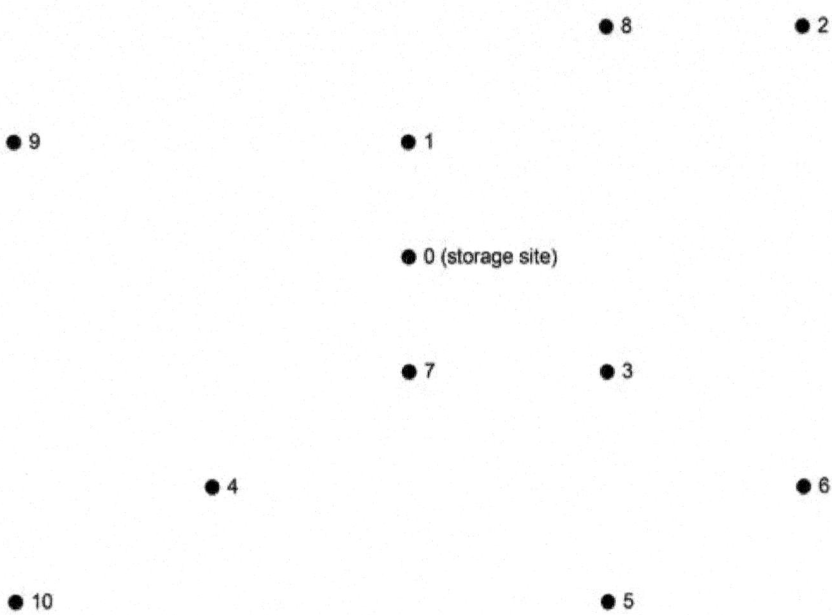

Figure 1: Topology of the problem.

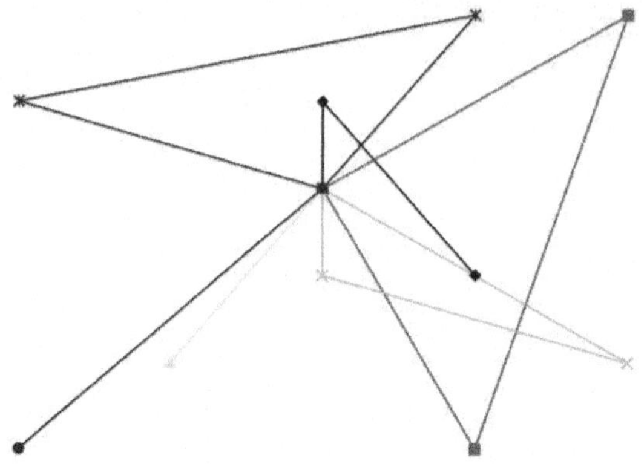

Figure 2: Generation 10, distance: 205.61.

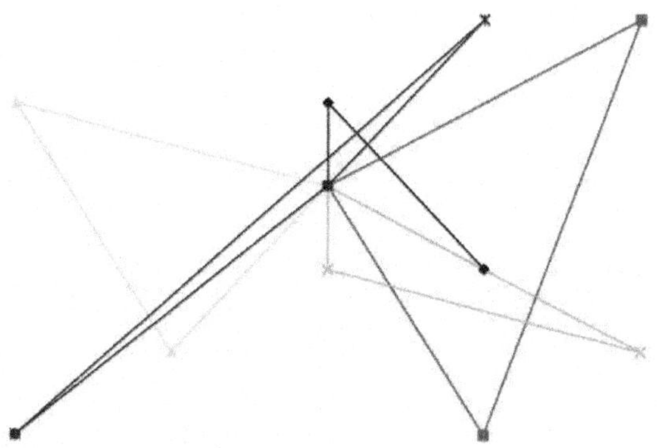

Figure 3: Generation 20, distance: 205.56.

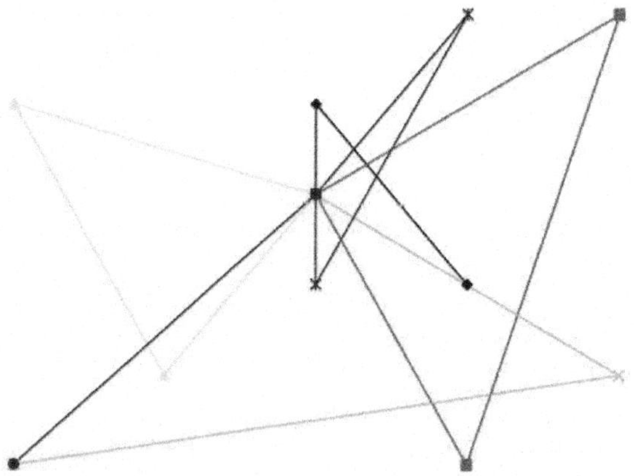

Figure 4: Generation 45, distance: 201.65.

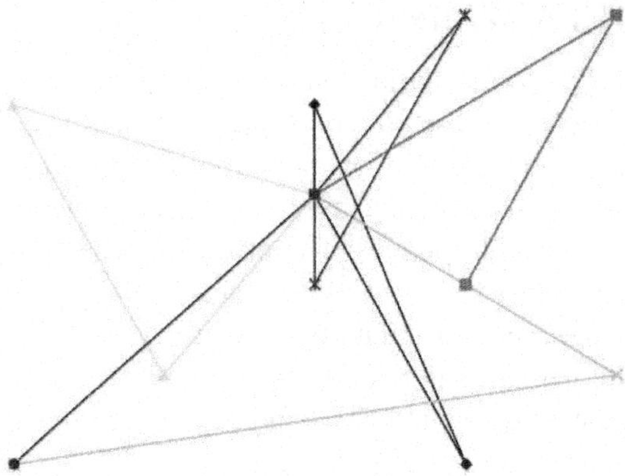

Figure 5: Generation 75, distance: 201.40.

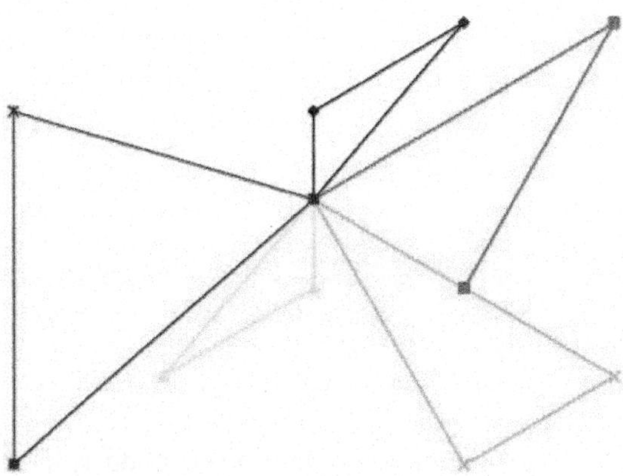

Figure 6: Generation 100, distance: 169.76.

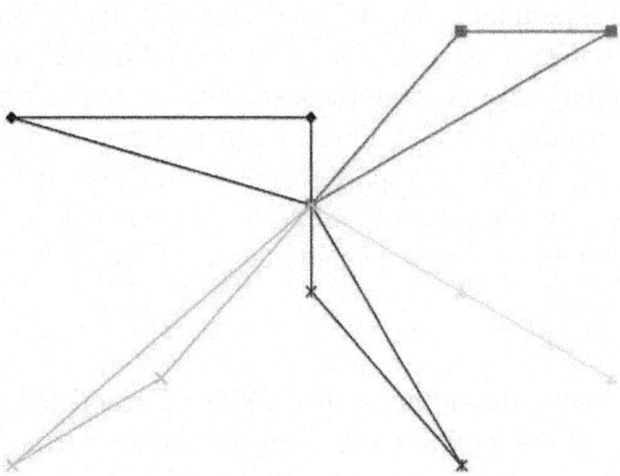

Figure 7: Generation 150, distance: 153.06.

EXPERIMENTS AND RESULTS

Van Breedam [25] presents a large number of instances of features and values. He suggested also a reduced set of representative cases to be solved by different heuristics. Some of them involve 100 nodes with a total demand of 1000 units. We follow here this convention and use the author's nomenclature for the cases with the same capacity of vehicles. The problems tested range from Bre-1 to Bre-15. From Bre-1 to Bre-6 only the capacity of vehicles is restricted. The demand on each client node is 10 units, while the vehicle capacity is 100 units in Bre-1 and Bre-2, 50 units in Bre-3 and Bre-4, while it is of 200 units in Bre-5 and Bre-6. Problems Bre-7 and Bre-8 have not been considered because they involve both deliveries and pickups.

Table 7: Results for the problems of Van Breedam [25,26].

Instance	Best Known Solution	Best Solution Found					
		Best Solution	Evaluations	Success (%)	Average Solution	Δ (%)	Average Time (min)
Bre-1	1106.00	1157.00	395347.12 ± 125177.89	42.00	1169.02 ± 14.76	−4.61	25.16
Bre-2	1506.00	1587.00	374623.02 ± 25544.61	18.00	1595.75 ± 16.91	−5.38	24.11
Bre-3	1751.00	1751.00	121640.67 ± 25684.27	100.00	1751.00 ± 0.00	0.00	11.20
Bre-4	1470.00	1470.00	287936.32 ±14739.13	100.00	1470.00 ± 0.00	0.00	19.68
Bre-5	950.00	985.00	326733.84 ± 112846.01	76.00	994.12 ± 11.49	-3.68	21.66
Bre-6	969.00	979.00	523967.47 ± 162502.07	62.00	982.43 ± 6.59	−1.03	31.72
Bre-9	1690.00	1790.00	657281.22 ± 21838.39	17.00	1797.93 ± 8.03	-5.92	38.52
Bre-10	1026.00	1026.00	871531.96 ± 326830.70.	52.00	1035.85 ± 9.42	0.00	49.45
Bre-11	1128.00	1187.00	1342635.29 ± 543765.65	100.00	1187.00 ± 0.00	−5.23	73.47

In most of these problems a single type of good is assumed. Exceptions are problems Bre-9 to Bre-11, in which demands are heterogeneous. Finally Bre-12 to Bre-15 are problems involving time windows, and therefore are not included in our experiments.**Table 7** presents the results of our tests. The best known solution, drawn from [25,26], is presented as well as information on the solutions under our procedure (best solution, average evaluation, success, mean solution, average running times and percentage of difference with the best known solution). With respect to the processing time, some instances took only minutes while others required hours, although in average the best solutions took not much longer than an hour to be found.

CONCLUSION

We presented a new algorithm for the CVRP (Capacitated Vehicle Routing Problem), distinguishing between the standard and its alternative formulation. The details of the procedure were described and the comparison with results in the literature has been presented. While the outcomes are similar to them, their running times are not. In experiments not discussed here we noticed that problems that exceed 100 nodes the processing time is much longer, due to the demands of the first stage of the algorithm. We plan to add further search filters at the optimization stage as well as to try with other meta-heuristics to run on the alternative model.

ACKNOWLEDGEMENTS

This work has been supported by several funding sources. The authors are in particular thankful to the National Research Council of Argentina (CONICET) as well as to the Universidad Nacional del Sur project PGI 24/J056.

REFERENCES

1. D. Ambrosino and A. Sciomachen, "A Food Distribution Network Problem: A Case Study," IMA Journal of Management Mathematics, Vol. 18, No. 1, 2007, pp. 33-53.doi:10.1093/imaman/dpl012
2. L. D. Bodin and B. L. Golden, "Classification in Vehicle Routing and Scheduling," Networks, Vol. 11. No. 2, 1981, pp. 97-108. doi:10.1002/net.3230110204
3. S. N. Kumar and R. Panneerselvam, "A Survey on the Vehicle Routing Problem and Its Variants," Intelligent Information Management, Vol. 4, No. 3, 2012, pp. 66-74.doi:10.4236/iim.2012.43010
4. G. Laporte and I. H. Osman, "Routing Problems: A Bibliography," Annals of Operations Research, Vol. 61, No. 1, 1995, pp. 227-262. doi:10.1007/BF02098290
5. E. Mota, V. Campos and A. Corbéran, "A New Metaheuristic for the Vehicle Routing Problem with Split Demands," In: J. van Hemert and C. Cotta, Eds., Evolutionary Computation in Combinatorial Optimization, Lecture Notes in Computer Science Vol. 4446, Springer-Verlag, Berlin, 2007, pp. 121-129. doi:10.1007/978-3-540-71615-0_11
6. R. Tavakkoli-Moghaddam, N. Safaei, M. Kah and M. Rabbani, "A New Capacitated Vehicle Routing Problem with Split Service for Minimizing Fleet Cost by Simulated Annealing," Journal of the Franklin Institute, Vol. 344, No. 5, 2007, pp. 406-425.doi:10.1016/j.jfranklin.2005.12.002
7. J. M. Belenguer, E. Benavent, N. Labadi, C. Prins and M. Reghioui, "Split Delivery Capacitated Arc-Routing Problem: Lower Bound and Metaheuristic," Transportation Science, Vol. 44, No. 2, 2010, pp. 206-220. doi:10.1287/trsc.1090.0305

8. S. R. Thangiah, A. Fergany and S. Awan, "Real-Time Split-Delivery Pickup and Delivery Time Window Problems with Transfers," Central European Journal of Operations Research, Vol. 15, No. 4, 2007, pp. 329-349. doi:10.1007/s10100-007-0035-x
9. M. Desrochers, J. Desrosiers and M. M. Solomon, "A New Optimization Algorithm for the Vehicle Routing Problem with Time Windows," Operations Research, Vol. 40, No. 2, 1002, pp. 342-354. doi:10.1287/opre.40.2.342
10. C. G. Lee, M. A. Epelman, C. White and Y. A. Bozer, "A Shortest Path Approach to the Multiple-Vehicle Routing Problem with Split Pick-Ups," Transportation Research, Vol. 40, No. 4, 2006, pp. 265-284. doi:10.1016/j.trb.2004.11.004
11. P. Belfiore and H. T. I. Yoshizaki, "Scatter Search for a Real-Life Heterogeneous Fleet Vehicle Routing Problem with Time Windows and Split Deliveries in Brazil," European Journal of Operational Research, Vol. 199, No. 3, 2009, pp. 750-758.doi:10.1016/j.ejor.2008.08.003
12. J. H. Wilck IV and T. M. Cavalier, "A Construction Heuristic for the Split Delivery Vehicle Routing," American Journal of Operations Research, Vol. 2, No. 2, 2012, pp. 153-162.doi:10.4236/ajor.2012.22018
13. Y. Nagao and H. Nagamochi, "A DP-Based Heuristic Algorithm for the Discrete Split Delivery Vehicle Routing Problem," Journal of Advanced Mechanical Design, Systems, and Manufacturing, Vol. 1, No. 2, 2007, pp. 217-226. doi:10.1299/jamdsm.1.217
14. Y. Yu, H. Chen and F. Chu, "A New Model and Hybrid Approach for Large Scale Inventory Routing Problems," European Journal of Operational Research, Vol. 189, No. 3, 2008, pp. 1022-1040. doi:10.1016/j.ejor.2007.02.061
15. M. A. Nowak, O. Ergun and C. C. White, "An Empirical Study on the Benefit of Split Loads with the Pickup and Delivery Problem," European Journal of Operational Research, Vol. 198, No. 3, 2009, pp. 734-740. doi:10.1016/j.ejor.2008.09.041
16. Z. Shen, M. M. Dessouky and F. Ordóñez, "A Two-Stage Vehicle Routing Model for Large-Scale Bioterrorism Emergencies," Networks, Vol. 54, No. 4, 2009, pp. 255-269.doi:10.1002/net.20337
17. L. Moreno, M. P. de Aragao and E. Uchoa, "Improved Lower Bounds for the Split Delivery Vehicle Routing Problem," Operations Research Letters, Vol. 38, No. 4, 2010, pp. 302-306. doi:10.1016/j.orl.2010.04.008
18. M.-C. Bolduc, G. Laporte, J. Renaud and F. F. Boctor, "A Tabu Search Heuristic for the Split Delivery Vehicle Routing Problem with Production

and Demand Calendars," European Journal of Operational Research, Vol. 202, No. 1, 2010, pp. 122-130.doi:10.1016/j.ejor.2009.05.008

19. U. Derigs, B. Li and U. Vogel, "Local Search-Based Metaheuristics for the Split Delivery Vehicle Routing Problem," Journal of the Operational Research Society, Vol. 61, No. 9, 2009, pp. 1356-1364. doi:10.1057/jors.2009.100
20. M. Jin, K. Liu and R. O. Bowden, "A Two-Stage Algorithm with Valid Inequalities for the Split Delivery Vehicle Routing Problem," International Journal of Production Economics, Vol. 105, No. 1, 2007, pp. 228-242. doi:10.1016/j.ijpe.2006.04.014
21. R. E. Aleman and R. R. Hill, "A Tabu Search with Vocabulary Building Approach for the Vehicle Routing Problem with Split Demands," International Journal of Metaheuristics, Vol. 1, No. 1, 2010, pp. 55-80. doi:10.1504/IJMHEUR.2010.033123
22. R. E. Aleman, X. Zhang and R. R. Hill, "An Adaptive Memory Algorithm for the Split Delivery Vehicle Routing Problem," Journal of Heuristics, Vol. 16, No. 3, 2010, pp. 441-473. doi:10.1007/s10732-008-9101-3
23. S. Basu, "Tabu Search Implementation on Traveling Salesman Problem and Its Variations: A Literature Survey," American Journal of Operations Research, Vol. 2, No. 2, 2012, pp. 163-173. doi:10.4236/ajor.2012.22019
24. D. Goldberg, "Genetic Algorithms in Search, Optimization, and Machine Learning," Addison-Wesley Longman Publishing Co., Inc., Boston, 1989.
25. A. Van Breedam, "An Analysis of the Behavior of Heuristics for the Vehicle Routing Problem for a Selection of Problems with Vehicle Related, Customer-Related, and Time-Related Constraints," Ph.D. Thesis, University of Antwerp—RUCA, Antwerpen, 1994.
26. E. Alba and B. Dorronsoro, "A Hybrid Cellular Genetic Algorithm for the Capacitated Vehicle Routing Problem," In: Engineering Evolutionary Intelligent Systems, Chapter 13, Studies in Computational Intelligence, SpringerVerlag, Heidelberg, 2008, pp. 379-422.doi:10.1007/978-3-540-75396-4_14

Chapter 3

GREEN INVESTMENT COST OPTIMIZATION MODEL IN THE SUPPLY CHAIN

Seungbae Sim[1], Hosang Jung[2]

[1]Center for Defense Acquisition, Korea Institute for Defense Analyses, Seoul, Republic of Korea

[2]Graduate School of Logistics, Inha University, Incheon, Republic of Korea

ABSTRACT

The objective of this study is to develop a model that determines the optimal points for investment in green management by defining a mathematical relationship between carbon trading profits and investments in green management using a company's supply chain information. To formulate this model, we first define and analyze a green supply chain in a multi-dimensional and quantitative manner. The green investment alternatives considering in our model are as follows: 1) purchasing eco-friendly raw materials that cost more than conventional raw materials but whose use in production results in lower CO_2 **emissions; 2) replacing current facilities with new eco-friendly facilities that have the capability to reduce CO_2 emissions; and 3) changing modes of transport from less eco-friendly to more eco-friendly modes. We propose a green investment cost optimization (GICO)** model that enables us to determine the optimal investment points. The proposed GICO model can support decision-making processes in green supply chain management environments.

INTRODUCTION

International debates about reduction of greenhouse gas emissions, such as the debate surrounding the Kyoto Protocol, are in progress, focusing on the participation of developed countries. In these debates, signatory countries are classified into two groups: obligatory participants and voluntary participants. Most of the obligatory participants are developed countries, such as the Organization for Economic Cooperation and Development (OECD) member countries, while most developing countries are considered voluntary participants.

International protocols usually set the reduction targets for greenhouse gas emissions for the developed signatory countries. According to the Kyoto Protocol, the target for each developed country is to reduce greenhouse gas emissions by an average of 5%. In the near future, the PostKyoto Protocol will require the developed signatory countries to reduce greenhouse gas emissions further.

These countries need to consider introducing nationwide legal systems to address international regulations. Nationwide greenhouse gas-related policies can address how to induce domestic industries and companies to reduce emissions. A national government can introduce a greenhouse gas emissions allocation system ("carbon taxes") that gives an allowance of greenhouse gas emissions to each firm or institutionalizes a carbon trading system. A number of countries, especially OECD countries, have implemented carbon tax systems for greenhouse gas emissions. In many cases, carbon taxes are levied on energy-consuming activities (e.g., electric power generation, operating facilities, and running machines) and motor vehicle activities (e.g., transportation and heavy equipment operation).

Along with a carbon tax system, a carbon trading system can be an effective method for controlling worldwide greenhouse gas emissions. In 2005, the EU implemented the EU ETS (emissions trading system), which is a European policy to cope with climate change and reduce industrial greenhouse gas emissions in a cost-effective manner. Because carbon trading gives companies incentives to reduce their greenhouse gas emissions, the use of carbon trading is expected to become more widespread throughout the world.

In this study, we assume that government regulations pertaining to greenhouse gas emission allowances are constantly being enhanced, which can contribute a huge amount to the cost.

From the perspective of a supply chain involving various firms, such as suppliers, manufacturers, distributors, and retailers, the cumulative greenhouse gas emissions per unit product increase as materials and products move from one stage to the next. A supplier's raw materials can include carbon, which will contribute to greenhouse gas emissions during production, and the transportation of those raw materials results in greenhouse gas emissions by transporting vehicles. A manufacturer can generate greenhouse gases during the production (e.g., operating machines) and also during transportation of products from the manufacturer to the distributor. The distributor maintains inventories, which also results in greenhouse gas emissions, albeit in an indirect way (e.g., using electricity to refrigerate foods). Transportation of products from the distributor to the retailer also generates greenhouse gases. Finally, consumers purchase products from the retailer and use them. Transportation of products from retailers to consumers generates greenhouse gases, as does the use of products by consumers, particularly the usage of electricity, which is mainly produced from fossil fuels. In short, as products move along a supply chain from supplier to consumer, the revenue of the firm increases but the cumulative emission of greenhouse gases increases as well.

The objective of this study is to develop a mathematical model to determine the optimal green investment cost in the supply chain, assuming that the allowance of greenhouse gas emissions for a firm will decrease due to expected increases in the stringency of regulations on carbon dioxide emissions. The alternatives for our green investment are as follows: 1) purchasing eco-friendly raw materials that cost more than convention raw materials but whose use in production results in lower CO_2 emissions; 2) replacing current facilities with new ecofriendly facilities that have the capability to reduce CO_2 emissions; and 3) changing modes of transport from less eco-friendly to more eco-friendly modes.

This paper is organized as follows. In Section 2, we review the literature on green supply chain management. In Section 3, we present a mathematical model for green investment cost optimization in the supply chain. In Section 4, we demonstrate the applicability of the proposed model using a numerical example. In Section 5, we present concluding remarks and suggestions for further research.

LITERATURE REVIEW

In this section, we review the literature on green supply chain management (GSCM) issues such as green purchasing, green supply chain planning, and green supply chain design. Reference [1] proposed a green supply chain planning model for aluminum processing processes to consider the environmental effects of the generated supply chain plan. Reference [2] developed a mixedinteger programming model for determining green investment for technology adoption and facility location in terms of both the required cost and the environmental impact. Reference [3] argued that "green purchasing", "green manufacturing/materials management", "green distribution/marketing" and "reverse logistics" could be regarded as the key elements of GSCM. Reference [4] proposed a credibility-based fuzzy mathematical programming model to address a green logistics network design problem under uncertain conditions and solved this model using an interactive fuzzy solution approach. The proposed model took into account both environmental and economic aspects in the design of the multi-stage logistics network of interest. The CO_2 equivalent index was used to model the environmental impact across the logistics network.

Various studies have been conducted on the validity and efficiency of the implementation of GSCM [5-8]. Reference [5] identified and evaluated hypotheses on eco-friendly purchasing strategies, and [6] examined the components of GSCM, their role, and relative importance through an analytical network process (ANP). Reference [7] conducted a study on the effects of eco-friendly purchasing in supply chains. Reference [8] examined the effects of GSCM from the perspective of managing environmental control systems. Reference [9] analyzed the relation between average SO_2 concentration and total environmental investment costs, such as pollution-preventing equipment cost, using a simulation model. The results showed that minimizing total environmental investment costs was not an adequate solution and that positive environmental investments were effective in the prevention of pollution and protection of a healthy natural environment.

In recent years, some studies on GSCM using direct modeling have been conducted. Reference [10] addressed the effort involved in maximizing profit while reducing carbon dioxide emissions in supply chains assuming the existence of a carbon tax and a carbon emissions trading system. Reference [11] developed and solved a mixed integer programming model for a carbon-sensitive supply chain that minimizes emissions throughout the supply chain by taking green procurement into consideration. The analysis results showed that companies could reduce their carbon emissions significantly with the introduction of carbon pricing by decentralizing the supply chain and multi-sourcing to reduce transportation and production emissions. As for sustainable supply chain management (SSCM), which could be regarded as a broader concept than GSCM, [12] summarized the related literature well for those who have an interest in both GSCM and SSCM.

Although green investment has been a key issue in the area of GSCM, there has been little research on the optimization of green investment. Accordingly, the next pressing issue in GSCM might be the optimization of green investment costs to reduce greenhouse gases. Thus, as mentioned in the previous section, a mathematical model for green investment cost optimization needs to be developed. Through optimized green investment, firms will be able to appropriately respond to fast-changing global environmental regulations or issues and strengthen their global competitiveness.

GREEN INVESTMENT COST OPTIMIZATION

We consider a supply chain network consisting of a supplier, a manufacturer, and a distributor. The supplier provides the manufacturer with raw materials, and the manufacturer produces a product by processing the raw materials or by assembling multiple components that might be made of various raw materials. Each component can be processed on a specific machine; for example, component c1 is only processed on machine m1 or m3. After the manufacturing processes, the final product moves to the distribution center or the consumer via transport modes such as road, rail, air, and marine.

Before proceeding to build a mathematical model, the following assumptions must be made:

A1: The firm owns its supply chain and has the power to control it.

A2: The firm's supply chain has sufficient production and transportation capacity.

A3: Greenhouse gases are generated along the whole supply chain (e.g., production and transportation).

A4: The firm is given a greenhouse gas emission allowance in accordance with related regulations.

A5: If the total emissions of the firm exceed the allowance, the government imposes a CO_2 tax.

A6: If the total emissions of the firm are lower than the allowance, the firm can sell the margin (emissions permit).

A7: Another firm that needs to obtain an emissions permit must buy it from the firm, which wants to sell it.

Raw materials purchased from a supplier have indirect greenhouse gas emissions associated with them which are generated by raw material production processing. In the manufacturing stage, raw materials are transformed into components or products. Greenhouse gases are emitted by chemical processes or assembly processes using electric power. Final products are distributed from a factory to a distribution center by various transport modes (e.g., road, railroad, and air). Each transportation system emits greenhouse gases in the use of fossil fuels or electric power.

Assumptions 4 and 5 indicate that the firm has to meet its allowance level of greenhouse gas emissions if possible. If the firm exceeds this level, a greenhouse gas tax can be charged by the authorities concerned. If the greenhouse gas emissions of the firm are below this level, the firm can sell the surplus on the carbon trading market.

Assumptions 6 and 7 indicate that emission's trading is a market-based approach used to control pollution by providing economic incentives for achieving reductions in CO_2 emissions [13].

We now propose a green investment cost optimization model (GICO model). A mathematical model is established for the cost optimization problem. Without loss of generality, the purchasing lead times and production setup times is not considered in this model. The decision variables of the model are the production quantity, the transportation quantity for each transport mode, the emissions reduction coefficients, raw material costs, manufacturing costs from component to product, transportation costs, investment costs for raw materials, facilities and so on. The tradable permit quantity of CO_2 and the excess CO_2 emissions quantity have either 0 or positive real values.

Indices

r : raw material;

c : component;

p : product;

m : machine;

l : transport mode;

t, t_0, T : fiscal year, base year, and target year.

BOM Parameters

$q_{r,c}$: amount of raw material r to product one unit of component c;

$q_{c,p}$: amount of component c to produce one unit of product p;

S_m : set of components which can be processed at machine m.

Cost (Profit) Parameters

$c_{tax,t}$: unit GHG tax in time period t;

$c_{trade,t}$: unit CO_2 trading profit in time period t;

$c_{penalty}$: penalty cost of unsatisfied GHG emission target quantities at target year.

Green Investment Parameters

α_t: GHG investment ratio in time period t;

$\alpha_{r,t}$: GHG emissions reduction per unit green investment cost of raw material r in time period t;

$\alpha_{c,m,t}$: GHG emission reduction quantities per unit green investment cost of (c, m) in time period t;

$\alpha_{p,l,t}$: GHG emission reduction quantities per unit green investment cost of (p, l) in time period t.

Decision Variables

$x_{p,t}$: amount of production of product p in time period t;

$y_{p,l,t}$: amount of transportation of product p by transport mode l in time period t;

A_t: actual GHG emission quantities in time period t;

I_t: GHG emission reduction quantity by green investment in time period t;

$z_{r,t}$: investment cost of raw material r in time period t;

$z_{c,m,t}$: investment cost of (c, m) in time period t;

$z_{p,l,t}$: investment cost of (p, l) in time period t;

β_t^- : tradable permit quantity in time period t;

β_t^+ : excess emissions quantity in time period t;

ε_t : unsatisfied emissions quantity in time period t;

b_t: green investment budget in time period t;

$e_{r,t}$: unit GHG emissions of raw material r purchased in time period t;

$e_{c,m,t}$: unit GHG emissions of component c when processed at machine m in time period t;

$e_{p,l,t}$: unit GHG emissions of product p when distributed by transport mode l in time period t;

$c_{r,t}$: unit cost of raw material r in time period t;

$c_{c,m,t}$: unit processing cost of product p by transport mode l in time period t;

$c_{p,l,t}$: unit transportation cost of product p by transport mode l in time period t;

$p_{p,t}$: unit profit of product p in time period t.

Constants

E_t : GHG emissions allowed in time t;

$C_{p,t}$: production capacity of product p in time period t;

$C_{l,t}$: transportation capacity of transport model l in time period t;

R_l : transportation load sharing ratio of transport mode l;

V_p : loading volume of product p;

$D_{p,t}$: demand of product p in time period t;

p : profit margin rate.

The GICO model can be formulated as follows:

Maximize

$$\sum_{p,t} p_{p,t} x_{p,t} - \sum_{t} c_{GHG,t} \beta_t + \sum_{t} c_{trade,t} \beta_t^{+} - c_{penalty} \varepsilon_T \qquad (1)$$

Subject to the following constraints:

$$A_t - I_t - E_t + \left(\beta_t^{+} - \beta_t^{-}\right) = 0, \forall t \qquad (2)$$

$$A_t = \sum_{r,c,p} q_{r,c} q_{c,p} x_{p,t} e_{r,t} + \sum_{c \in S_m, m, p} q_{c,p} x_{p,t} e_{c,m,t} + \sum_{p,l} y_{p,l,t} e_{p,l,t}, \forall t \quad (3)$$

$$I_t = \sum_r \alpha_{r,t} z_{r,t} + \sum_{c \in S_m, m} \alpha_{c,m,t} z_{c,m,t} + \sum_{p,l} \alpha_{p,l,t} z_{p,l,t}, \forall t \quad (4)$$

$$A_t - I_t - \beta_t^- \leq E_t, \forall t \quad (5)$$

$$A_t - I_t + \beta_t^+ \geq E_t, \forall t \quad (6)$$

$$A_t \geq I_t, \forall t \quad (7)$$

$$A_T - \varepsilon_T \leq E_T \quad (8)$$

$$c_{r,t+1} = c_{r,t} + \frac{z_{r,t}}{\sum_p q_{r,c} q_{c,p} x_{p,t}}, \forall r, t \quad (9)$$

$$c_{r,t+1} \geq c_{r,t}, \forall r, t \quad (10)$$

$$e_{r,t+1} = e_{r,t} - \frac{\alpha_{r,t} z_{r,t}}{\sum_p q_{r,c} q_{c,p} x_{p,t}}, \forall r, t \quad (11)$$

$$e_{r,t+1} \leq e_{r,t}, \forall r, t \quad (12)$$

$$c_{c,m,t+1} = c_{c,m,t} + \frac{z_{c,m,t}}{\sum_p q_{c,p} x_{p,t}}, \forall c, m, t \quad (13)$$

$$c_{c,m,t+1} \geq c_{c,m,t}, \forall c, m, t \quad (14)$$

$$e_{c,m,t+1} = e_{c,m,t} - \frac{\alpha_{c,m,t} z_{c,m,t}}{\sum_p q_{c,p} x_{p,t}}, \forall c, m, t \quad (15)$$

$$e_{c,m,t+1} \leq e_{c,m,t}, \forall c,m,t \tag{16}$$

$$\sum_{p} c_{p,l,t+1} = \sum_{p} c_{p,l,t} + \sum_{p} z_{p,l,t}, \forall l,t \tag{17}$$

$$c_{p,l,t+1} \geq c_{p,l,t}, \forall p,l,t \tag{18}$$

$$\sum_{p} e_{p,l,t+1} = \sum_{p} e_{p,l,t} - \sum_{p} \alpha_{p,l,t} z_{p,l,t}, \forall l,t \tag{19}$$

$$e_{p,l,t+1} \leq e_{p,l,t}, \forall p,l,t \tag{20}$$

$$\sum_{r} z_{r,t} + \sum_{c \in S_m, m} z_{c,m,t} + \sum_{p.l} z_{p,l,t} \leq b_t, \forall t \tag{21}$$

$$b_{t+1} = \alpha_t \sum_{p} p_{p,t} x_{p,t},\ t > t_0 \tag{22}$$

$$p_{p,t} = p\left\{ \sum_{r,c} q_{r,c} q_{c,p} c_{r,t} + \sum_{c \in S_m, m} q_{c,p} c_{c,m,t} + \sum_{l} R_l c_{p,l,t} \right\},$$
$$\forall p,t \tag{23}$$

$$x_{p,t} = D_{p,t}, \forall p,t \tag{24}$$

$$x_{p,t} = \sum_{l} y_{p,l,t}, \forall p,t \tag{25}$$

$$\sum_{p} y_{p,l,t} = R_l \sum_{p,l} y_{p,l,t}, \forall l,t \tag{26}$$

$$x_{p,t} \leq C_{p,t}, \forall p,t \tag{27}$$

$$\sum_{p} V_p y_{p,l,t} \leq C_{l,t}, \forall l,t \tag{28}$$

$$x_{p,t} \geq 0, p_{p,t} \geq 0, \forall p,t \tag{29}$$

$$y_{p,l,t} \geq 0, \forall p,l,t \tag{30}$$

$$z_{r,t} \geq 0, e_{r,t} \geq 0, c_{r,t} \geq 0, \forall r,t \tag{31}$$

$$z_{c,m,t} \geq 0, e_{c,m,t} \geq 0, c_{c,m,t} \geq 0, \forall c, m, t \tag{32}$$

$$z_{p,l,t} \geq 0, e_{p,l,t} \geq 0, c_{p,l,t} \geq 0, \forall p, l, t \tag{33}$$

$$A_t \geq 0, I_t \geq 0, b_t \geq 0, \beta_t^+ \geq 0, \beta_t \geq 0, \forall t \tag{34}$$

Equation (1) is the objective function of the GICO model, which includes the total greenhouse gas tax, the carbon trading profit and the penalty cost. In the target year, the firm must meet a greenhouse gas emissions target set by government or the firm itself. If the firm does not meet this goal, the greenhouse gas penalty cost, which is the product of the unit penalty cost and unsatisfied emissions quantities, can be charged. For example, the greenhouse gas penalty cost is \$1,000,000 when the unit penalty cost is \$1000 per tCO_2 and the unsatisfied emissions quantity is 1000 tCO_2. This objective function minimizes the total cost of greenhouse gas. The constraints of the model are given as Equations (2) to (34).

Equation (2) is the CO_2 emissions balance equation, according to which current actual CO_2emission quantities minus reduction quantities through green investment minus greenhouse emissions allowed plus tradable emission quantities minus excess emission quantities equals zero. Equations (3) and (4) define total CO_2 emissions quantities accumulated from raw material to final product in the supply chain and the green investment cost of raw materials or manufacturing and transportation facilities. Equations (5) and (6) define mutually exclusive constraints for CO_2emissions excess or surplus. If the tradable permit quantity is positive, the excess emissions quantity is zero. Similarly, if the excess emissions quantity is positive, the tradable permit quantity is zero.

Equation (7) implies that the reduction quantities achieved by green investment are at most the current emission quantities.

We assume that the actual CO_2 emission quantities exceed CO_2 emissions allowed in time period t_0. As mentioned earlier with respect to Assumption 4, this implies that the firm needs to reduce CO_2 emissions in the supply chain in accordance with the related regulations. The alternatives for reducing CO_2 emissions are bearing a greenhouse gas tax burden or CO_2 trading.

The firm must meet the CO_2 reduction target by time t_T. In other words, the firm must reduce CO_2emissions in the supply chain from the base year t_0 to the target year t_T. Though the firm is obliged to meet the CO_2 reduction target identified by government regulations, ever-increasing demand for products will increase total CO_2 emissions. This means that the firm may not meet the

CO_2reduction goal. In view of this situation, we assume an unsatisfied emissions quantity in target year T and add the related penalty cost from Equation (1). Equation (8) describes the CO_2 reduction target constraint at time t_T.

Equations (9) and (10) imply that the unit cost of raw material r in time period t + 1 increases by the green investment of raw material r divided by total raw material quantities purchased in time period t compared to the previous year. The GICO model focuses on green investment decision-making ensuring minimum economic profit and minimizing total CO_2 cost. Thus, we can assume production quantities equal to production demand, according to Equation (24). Equations (11) and (12) imply that the unit CO_2 emissions coefficient of raw material r in time period t + 1 decreases by the CO_2 emission reduction quantities divided by total raw material quantities purchased in time period t. This signifies return on green investment, which results in the use of eco-friendly raw material with lower CO_2 emissions than conventional materials.

Equations (13) and (14) imply that the unit cost of processing component c with machine m in time period t + 1 increases by the green investment of machine m divided by the total number of components, compared to previous year. Equations (15) and (16) imply that the unit CO_2emissions coefficient of processing and assembling component c with machine m in time period t + 1 decreases by CO_2 emission reduction quantities divided by the total number of components processed with machine m in time period t. As with Equations (11) and (12), these equations mean that green investment in processing facilities results in eco-friendly facilities with low CO_2 emission rates.

Equations (17) and (18) imply that total transportation costs by transport mode l in time period t + 1 are equal to total transportation costs by transport mode l in time period t plus the investment cost of transportation facilities in time period t. Equations (19) and (20) imply that the unit CO_2emissions coefficient of transportation facilities in time period t + 1 decreases by the CO_2 emission reduction quantities from green investment in transportation facilities in time period t.

Equation (21) is a budget constraint, and Equation (22) implies that the total budget in time period t + 1 is determined by the product of the greenhouse gas investment ratio and the total profit. For example, if the total profit is $1,000,000 and the greenhouse gas investment ratio is 5%, the total budget available during time period t+1 is $50,000. From Equation (23), the total profit is calculated as the fixed ratio to the total cost. We assume that the profit margin rate is 10%.

Equation (24) is the demand constraint for each product. As mentioned previously because the proposed model focuses on green investment decision-making for minimizing total CO_2 cost, we assume production quantities equal to production demand to evaluate the impact on green investment only.

Equation (25) is the transportation network constraint for each product, and Equation (26) is the transportation mix constraint for each transport mode. The transportation load sharing ratio of each transport mode in Equation (26) can be determined using national transportation statistics data.

Equations (27) and (28) are production and transportation capacity constraints, respectively. Transportation capacity is dependent on the product of the product volume and production quantities. Equations (29)-(34) are nonnegativity constraints.

The solution of the model yields green investment values for various alternatives, and these result in the replacement of raw materials and facilities by new ones. The value of greenhouse gas emissions at time t, namely current total emissions GHG_t, is expressed by Equation (35).

$$\begin{aligned} GHG_t &= A_t - I_t \\ &= \sum_{r,c,p} q_{r,c} q_{c,p} x^*_{p,t} e_{r,t} + \sum_{c \in S_m, m, p} q_{c,p} x^*_{p,t} e_{c,m,t} \\ &\quad + \sum_{p,l} y^*_{p,l,t} e_{p,l,t} - \sum_{r} \alpha_{r,t} z^*_{r,t} - \sum_{c \in S_m, m} \alpha_{c,m,t} z^*_{c,m,t} \end{aligned} \quad (35)$$

We introduce the concept of a Green Index (GI) to evaluate the current status in comparison with the base status and target status, using Equation (36). GI has a value between 0 and 1. The GI equals 1 when the greenhouse gas emissions target is met.

$$GI_t = \frac{GHG_{t_0} - GHG_t}{GHG_{t_0} - GHG_T}, \forall t \quad (36)$$

ILLUSTRATIVE EXAMPLE

To demonstrate the applicability of the proposed model, a case study based on a real situation in South Korea is presented. Though South Korea is not an obligatory participant in the Kyoto Protocol, the South Korean government aggressively set the reduction target for nationwide greenhouse gas emission at 4% below the 2005 level by 2020 to prepare for the Post-Kyoto mechanism, which will require the active participation of developing countries. This reduction target corresponds with 30% below BAU (business as usual). The BAU refers to the predicted level of greenhouse gas emissions that will be reached by a certain year if emissions increase at the current pace.

To meet the international requirements for greenhouse gas emissions reduction, the South Korean government is considering launching greenhouse gas emissions allocation and controlling systems. To address the Post-Kyoto mechanism, the government will expedite implementation of a carbon tax system and a carbon trading system. The South Korean government will introduce a capand-trade system for carbon dioxide in 2015. However, the greenhouse gas reduction plan still faces differences of opinion among policymakers and businesses concerning the feasibility of greenhouse gas emissions reduction by target year. Nevertheless, many companies in Korea will consider alternatives to greenhouse gas emissions reduction ranging from tax burdens to green investment.

We assume that there is an electronics company (Company A) in Korea that produces four types of products, namely, DRAM/flash memory, mobile phones, LCDs, and TVs. Company A has three factories, three distribution centers, and nine submarkets in the domestic market. The production and distribution facilities can be divided into three areas, namely, a metropolitan area, a southwestern area, and a southeastern area. The metropolitan area contains factory 1 and distribution center 1. The final products produced at factory 1 are gathered at distribution center 1. Products stored at distribution center 1 can be directly distributed into submarket 6 by truck or distributed into distribution center 2 and subsequently to submarket 7 by rail. We also assume that the following three transport modes exist: road, railroad, and air. In terms of transportation cost, road transportation is the most efficient of the three transport modes. In terms of carbon emissions, air transportation may be the worst of the three transport modes. We consider real situations and assume that the more expensive a transport mode is, the lower its carbon emission rate is, except for air transportation.

From the financial and sustainability report of Company A, we can calculate and estimate values of the costrelated parameters and CO_2 emissions-related parameters [14]. We extract domestic revenue and cost data from globally integrated data and domestic market share data. **Table 1**shows the supply chain information of Company A, which consists of total revenue and unit cost for each product type and the CO_2 emissions from raw material to customer. We assume that the total cost of each product is 70% of the total revenue and that total sales are calculated using estimated unit costs and total costs.

For example, the total revenue, total cost, total sales, total CO_2 (unit: tCO_2), and unit CO_2 (unit: tCO_2) for memory (P1) are 3880, 2716, 271,607,000, 1,776,757, and 0.0065, respectively. Similar to the memory, the revenue, cost, sales, and CO_2 values are generated for the other three products: mobile (P2), LCD (P3), and TV (P4).

Table 1: Supply chain information of company A

Product Type	Total Revenue (M$)	Total Cost (M$)	Total Sales (# of product)	Total CO_2 (tCO_2)	Unit CO_2 (tCO_2)
P_1	3880	2716	271,607,000	1,776,757	0.0065
P_2	8602	6021	30,107,805	3,939,092	0.1308
P_3	3654	2558	34,110,533	1,673,543	0.0491
P_4	5602	3921	3,921,960	2,565,608	0.6542

Then, we can calculate and estimate the unit raw material cost, unit assembly cost, and unit transportation cost of each transport mode. The raw material cost and assembly cost are calculated by multiplying the unit production cost by 30%. The cost of road transportation is calculated by multiplying the unit production cost by 40%. We assume that the cost of rail transportation is 150% of the road transportation cost and that the cost of air transportation is 200% of the road transportation cost. Air transportation is the most expensive of the three transport modes but is the worst transport mode in terms of carbon emissions.

Similarly, we can calculate and estimate unit CO_2 emissions coefficients for raw material costs, assembly, and transportation. The unit CO_2 emissions coefficient of P_4 is greater than that for any other products, but the total CO_2 emissions quantity for P_2 are greater than for that of P_4. To comply with the greenhouse gas emissions regulations, therefore, the company must control and reduce these coefficient values by priority.

Cost parameters, except c_{GHG} and c_{trade} and CO_2 emissions-related parameters, are generated using publicly available information about Company A. For example, total revenues for the company's four products and domestic market share data are used to generate cost parameters.

To focus on the impact of cost and CO_2 emissions on green investment, we apply a simple BOM and logistic structure in which one product is composed of one component, one component is composed of one raw material, and four products are manufactured in one factory and subsequently, distributed through one distribution center. We choose $15 per tCO_2 for c_{GHG}, which is regarded as a typical carbon emission tax rate, and $20 per tCO_2 for c_{trade} (Metcalf, 2007). The initial CO_2 reduction rates per unit green investment cost are set using the greenhouse gas reduction results for 2011. In 2011, Company A invested $360,700,000 in greenhouse gas reduction facilities and $342,300,000 in eco-friendly operations, which are different from traditional eco-unfriendly operations. As a result, the company was able to reduce greenhouse gas emissions by 346,000 tCO_2. With this information about CO_2 reduction, we can determine the CO_2 emissions reduction per unit green investment cost of raw material, assembly, and transportation. For example, the CO_2 emissions reduction coefficient of raw material is 0.0001477 tCO_2/$, which implies that if the company purchases eco-friendly raw material for \$1,000,000 instead of conventional raw material for \$5,000,000, the company can reduce CO_2 emissions quantities equivalent to the extra price. Thus, the CO_2 emissions reduction quantity is 0.0001477 tCO_2/\$ times \$5,000,000.

To verify the GICO model, we use the MILP (mixed integer linear programming) module of the ILOG OPL Development Studio 14.0 software tool. The basic model is based on the current situation in which product demand is ever increasing but regulation of CO_2 emissions is becoming stricter. We assume that demand increases at a rate of 5% per year and also assume that maximum CO_2emissions quantities allowed are decreasing at a rate of 5% per year.

The basic model results are as shown in **Figure 1** and can be analyzed in both economic and environmental terms. First, actual CO_2 emission quantities increase when demand increases and maximum CO_2 emission quantities allowed decrease. Thus, excess emission quantities in time period t increased by the corresponding amount, i.e., the gap between actual CO_2 emissions and maximum CO_2 emissions allowed. Second, the company cannot invest in eco-friendly raw materials, facilities, and transportation infrastructure because the earned value from green investment is less than the investment amount. Based on these analysis results, we can appreciate why the company hesitates about green investment. Third, the company cannot achieve the CO_2emissions reduction goal for the target year. Increasing demand and more stringent CO_2 reduction policies compel the company to accept CO_2 tax burdens and penalty costs instead of making green investments.

CONCLUSIONS

The objective of this study is to develop a mathematical model to determine the optimal green investment cost in a supply chain environment in which a company's allowable greenhouse gas emissions are expected to decrease because of increasingly stringent restrictions on carbon dioxide emissions. Our mathematical model considers the several alternatives for green investment, and the model is verified by the results of a case study.

The current study has several limitations. First, although the case study involved analysis of real company data, the company's detailed production plans and logistic network are not considered in the experimental design. Second, the assumptions about production environments also limit the study because inventory holding costs and operational electricity usage in the supply chain are not considered separately.

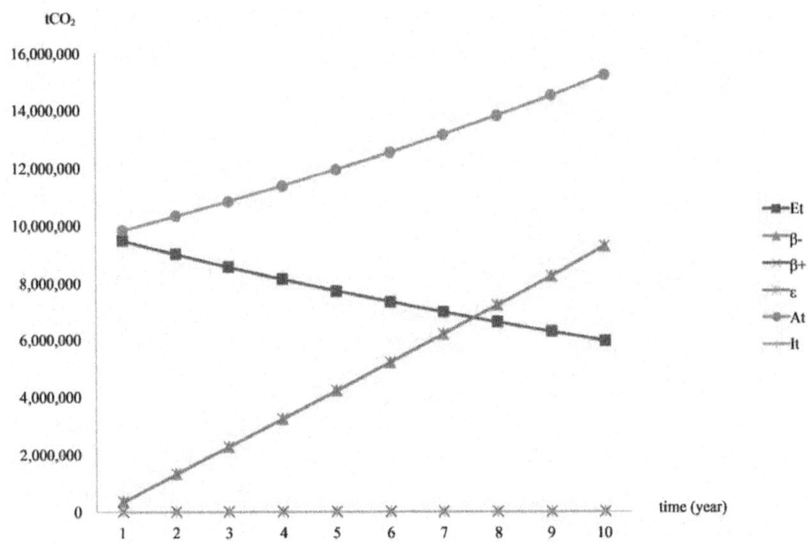

Figure 1: Basic model analysis results.

Third, the lack of greenhouse gas emissions reduction figures for new raw materials or new facilities limits the model to being based on publicly available statistical data.

As nationwide regulations on greenhouse gas emissions are enhanced, greenhouse gas emissions allowances will decrease. Therefore, companies will decide to invest at different points to reduce greenhouse gas emissions by means of environmentally friendly process improvements or will buy carbon emission permits from the carbon trading market. Nevertheless, the gap between actual CO_2 emissions and maximum CO_2 emissions allowed exists, and this gap cannot help but cause difficulties for companies in investing in eco-friendly infrastructure. In other words, companies must reduce total production quantities to meet the target level of greenhouse gas emissions.

The proposed GICO model can be used as a decision-making tool and as a negotiation tool between government and industries. It can assist companies in assessing the feasibility of green investment and can assist policymakers in assessing the effectiveness of greenhouse gas regulations.

REFERENCES

1. I. Ferretti, S. Zanoni, L. Zavanella and A. Diana, "Greening the Aluminum Supply Chain," International Journal of Production Economics, Vol. 108, No. 1-2, 2007, pp. 236-245.http://dx.doi.org/10.1016/j.ijpe.2006.12.037

2. A. Hugo and E. N. Pistikopoulos, "Environmentally Conscious Long-Range Planning and Design of Supply Chain Networks," Journal of Cleaner Production, Vol. 13, No. 15, 2005, pp. 1471-1491. http://dx.doi.org/10.1016/j.jclepro.2005.04.011
3. A. A. Hervani and M. M. Helms, "Performance Measurement for Green Supply Chain Management," Benchmarking: An International Journal, Vol. 12, No. 4, 2005, pp. 330-353.
4. M. S. Pishvaee, S. A. Torabi and J. Razmi, "CredibilityBased Fuzzy Mathematical Programming Model for Green Logistics Design under Uncertainty," Computers & Industrial Engineering, Vol. 62, No. 2, 2012, pp. 624-632.http://dx.doi.org/10.1016/j.cie.2011.11.028
5. H. M. William and P. Galle, "Green Purchasing Practices of US Firms," International Journal of Operations & Production Management, Vol. 21, No. 9, 2001, pp. 1222- 1238.http://dx.doi.org/10.1108/EUM0000000005923
6. J. Sarkis, "A Strategic Decision Framework for Green Supply Chain Management," Journal of Cleaner Production, Vol. 11, No. 4, 2003, pp. 397-409.http://dx.doi.org/10.1016/S0959-6526(02)00062-8
7. K. Green, B. Morton and S. New, "Green Purchasing and Supply Policies: Do They Improve Companies Environmental Performance?" Supply Chain Management, Vol. 3, No. 2, 1998, pp. 89-95. http://dx.doi.org/10.1108/13598549810215405
8. N. Darnall, G. J. Jolley and R. Handfield, "Environmental Management Systems and Green Supply Chain Management: Complements for Sustainability?" Business Strategy and The Environment Business Strategy Environment, Vol. 18, No. 1, 2008, pp. 30-45.http://dx.doi.org/10.1002/bse.557
9. M. Iijima, Y. Takemoro, Y. Oka, H. Kito, Y. Nishigaki, K. Kataoka and S. Asahi, "Economic Aspects of Environmental Investment in Plant Facilities," Computers & Industrial Engineering, Vol. 31, No. 3-4, 1996, pp. 713- 717. http://dx.doi.org/10.1016/S0360-8352(96)00244-6
10. H. Jung and S. J. Jeong, "A Green Supply Chain Planning Model Considering Carbon Taxes and Carbon Emission Trading," Journal of Korean Society of Supply Chain Management, Vol. 10, No. 1, 2010, pp. 13-22.
11. T. Abdallah, A. Diabat and D. Simchi-Levi, "A Carbon Sensitive Supply Chain Network Problem with Green Procurement," Proceedings of the 40th International Conference on Computers and Industrial

Engineering, Awaji, 25-28 July 2010, pp. 1-6.http://dx.doi.org/10.1109/ICCIE.2010.5668278

12. S. Seuring and M. Muller, "From a Literature Review to a Conceptual Framework for Sustainable Supply Chain Management," Journal of Cleaner Production, Vol. 16, No. 15, 2008, pp. 1699-1710. http://dx.doi.org/10.1016/j.jclepro.2008.04.020
13. G. E. Metcalf, "A Proposal for a US Carbon Tax Swap: An Equitable Tax Reform to Address Global Climate Change," The Hamilton Project, Brookings Institution, Washington DC, 2007.
14. Samsung, "2012 Sustainability Report," Samsung Group, Seoul, 2012.

Chapter 4

MONOTONIC VECTOR SPACE MODEL (I): CONCEPTS AND OPERATIONS

Jianwen Hu

National University of Defense Technology, Changsha, China

ABSTRACT

Monotonic vector space (MVS), as a novel model in which there exist some monotonic mappings, is proposed. MVS model is an abstract of many practical problems (such as image processing, system capability engineering, etc.) and includes many useful important operations. This paper, as the first one of series papers, discusses the MVS framework, relative important concepts and important operations including partition, synthesis, screening, sampling, etc. And algorithms for these operations are the focus of this paper. The application of these operations in system capability engineering will be dealt with in the second part of this series of papers.

INTRODUCTION

MVS is a special vector space where there exist some monotonic mappings. This model actually represents a lot of practical problems. Several examples are given as follows: Example 1 (Figure 1). Electric field intensity of point charge can be discussed in a vector space $S(x,y,z,t)$ which includes onetime dimension and three space dimensions x, y, and z, respectively on X, Y, and Z axis of three dimensional coordinate system (see Figure 1). Electric charge quantity of the point charge is also assumed to decrease with time increasing. So, it can be easily understood that there exists monotonic function relation between the electric field intensity $P(x,y,z,t)$ of point charge and the arbitrary one of the above four indices. This means $P(x,y,z,t)$ will monotonically increase or decrease with one of four indices changing monotonically while the others maintaining to be constants. Just in this sense, $S(x,y,z,t)$ is called monotonic vector space.

Example 2 (Figure 2) gives typical electric field intensity lobe of radar antenna. And the electric field intensity of arbitrary point $P(r,\theta)$ in the main lobe can be calculated in a vector space $V(r,\theta)$ which includes polar radius r and polar angle θ. Obviously, $P(r,\theta)$ monotonically increases or decreases with each of two parameters changing monotonically while the other maintaining to be constant. So, $V(r,\theta)$ can also be regarded as a MVS.

Example 3. A monotonic vector space P for a ground-to-air missile system may include the searching radar detecting capability metric dimension P_1, delay time metric dimension P_2, and tracking radar detecting capability metric dimension P_3. So, the equation $u=f_u(p_1, p_2, p_3)$ describes the relation between the three system capability metrics (i.e. P_1, P_2, and P_3) and one requirement metric (i.e. the kill probability index u). Function f_u is obviously monotonic, that is to say, the value of u (the kill probability) will monotonically increase or decrease with one of three system capability metric changing monotonically while the others maintaining to be constants. We call p_1 and p_3 monotonic increasing metric and p_2 monotonic decreasing metric.

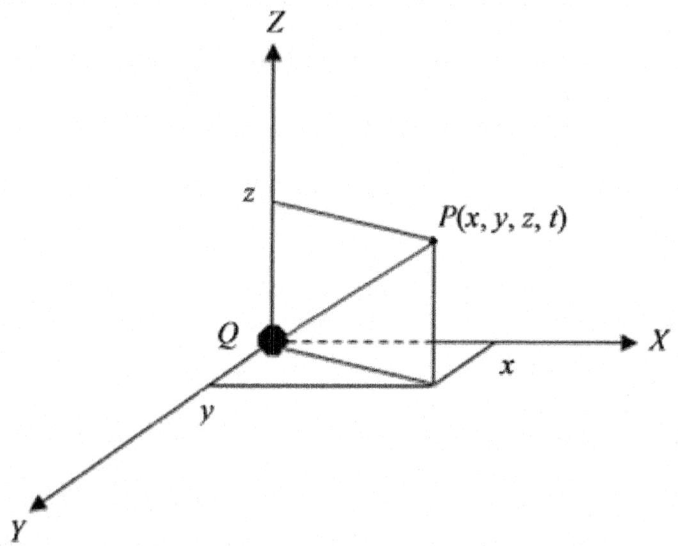

Figure 1: Electric field intensity of point charge.

Many practical problems can be solved with the operations defined in MVS. For example, for a vector space composed of system capability metrics [1,2], the requirement analysis of capability can be done with partitioning operation in MVS [2]. The screening operation in MVS may be applied for sensibility analysis of system capability metrics. Because Latin Hypercubic sampling [3] operation in MVS reduces variance largely, it is of great value to simulate experiments in system capability engineering.

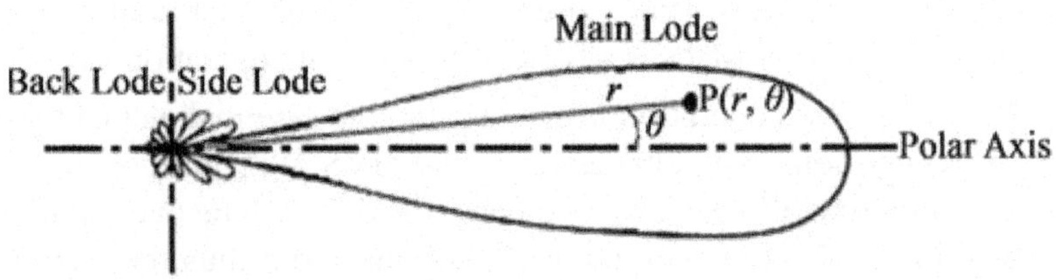

Figure 2: Typical electric field intensity lobe of radar antenna.

MONOTONIC VECTOR SPACE

Definition 1: Monotonic Vector Space (MVS) P. Let $P \subset R^n$, P is a point set, i.e., a n-dimensional subspace of R^n (R^n is n-dimensional Euclid space). There exist mappings such as $f_u : P \rightarrow u$ (u is a real variable and increases or decreases monotonically with each dimension of P). Every dimension of P can be continuous or discrete type. In most cases, it is very difficult to get the explicit analytic expression of f_u, which usually represents the corresponding relation between input and output of a black box. But many non-analytic methods (such as simulation method, numerical differentiation equation method, etc.) can be used to get the value of u (output). f_u can be classified into deterministic type and stochastic type. If f_u is of stochastic type, then $u = f_u(P) + \varepsilon$, ε is a random variable and $E(\varepsilon) = 0$.

According to the different type combination of dimension and mapping, MVS can be classified into four types, i.e. discrete deterministic type, continuous deterministic type, discrete stochastic type, and continuous stochastic type. The MVS researching framework is shown in Figure 3. Since many practical problems can be solved with the operations defined in MVS, mentioned as above, the discussion about operations related to MVS will be the focus of this paper. These operations usually include partitioning, synthesizing, screening, sampling, searching, and so on.

- *Partitioning.* This operation is used to attain the effective area, which covers all the points satisfying certain requirement, through segmenting MVS according to certain requirement criteria. For example, the effective detecting range of a jammed radar, the effective capability indicators fields, and so on, can be gotten through carrying out partitioning operation in MVS.

- *Synthesizing.* This operation is used to get the synthesized area of several effective areas, each of which contains points meeting different requirement targets. For example, acquiring the intersection of several effective areas in MVS is a typical synthesizing operation.
- *Screening.* This operation is used to find the essential elements such as key dimensions, key subarea, etc, in the MVS. With this operation, some less important dimensions or subarea of MVS can be overlooked in the tackling process of the practical problems. And thus the computation burden of other operations in MVS can be decreased remarkably.

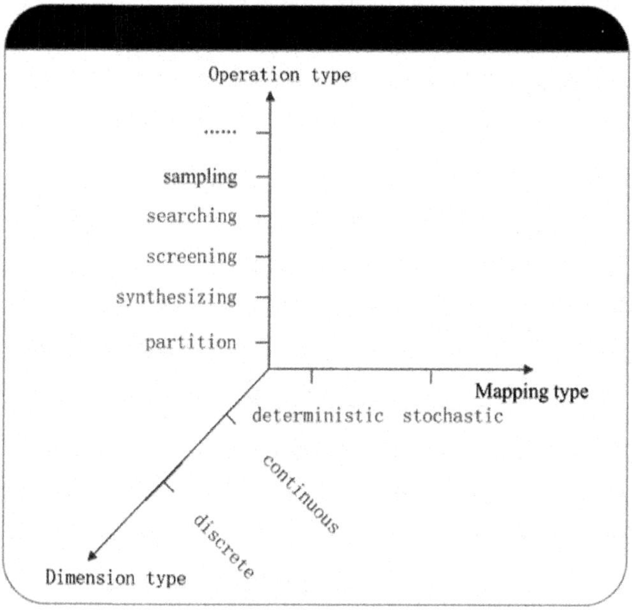

Figure 3: MVS researching frame.

- Sampling. If the mapping in MVS is of stochastic type, sampling the vector points in a large-scale MVS with high efficiency and quality have significance on stochastic simulation experiment (might be considered as stochastic mapping) in MVS.
- Searching. This operation is to find high sensitivity region for some mappings in MVS.
- Other operations. There are some other operations except for the above mentioned, such as slicing, projecting, rotating, etc. These operations, similar to OLAP or data mining, are useful for getting interesting information in the high dimensional MVS.

To solve practical problems, some new operations (except for those mentioned above) in MVS may need to be studied. In the following sections of this paper, the algorithms for partitioning operation in continuous/discrete-type MVS, synthesizing operation in continuous-deterministic-type MVS, sampling operation in stochastic-type MVS, screening operation and Searching operation in corresponding MVS, and so on, will be dealt with in detail as the following sections.

The Algorithm for Partition Operation in MVS of Continuous Deterministic Type

Partitioning is a very valuable operation in MVS and can be used to solve many practical problems such as capability requirement analysis, imaging processing, multi-objective optimization, etc. The requirement-based partitioning operation in MVS of continuous deterministic type is to be discussed in this section.

In the mapping $f_u : P \to u$, u_k is a specified requirement value and f_u is of deterministic type. With the requirement-based partitioning operation, MVS will be divided into two parts, i.e., the requirement-meeting part (zone) and the requirement-not-meeting part (zone). The requirement-meeting zone is called monotonic vector requirement locus (MVRL) in this paper. MVRL is critical for system capability analysis and design.

For simplicity, all the dimensions of MVS are assumed to be monotonically increasing (monotonically decreasing dimension can also be easily transformed into monotonically increasing dimension). MVRL is the set of points meeting $f_u(p_1, p_2, \ldots, p_n) \leq u_k, (\text{or} > u_k), u_k \geq 0$. And the value field of each dimension P_i is transformed into interval [0,Vi].

Some Definitions and Theorems

According to f_u and requirement constraints, MVRL can be produced through finding out all the requirementmeeting points of MVS. Since it is usually very difficult to get the analytic function of f_u, MVRL-acquiring method based on f_u without analytic expression, therefore, will be the discussion emphasis in this paper.

Definition 2: The minimal envelop hyperbox (MEH). MEH refers to the cross product of each dimension's interval [lower bound, upper bound] and it may also be represented as such simplified form

$$[0,V_1']\times[0,V_2']\times\cdots\times[0,V_n'](V_i'\leq V_i).$$

For example, assuming that x, y are the indexes of a system and that the

MVRL is the point set meeting the inequation $x^2+y^2\le 4(x\ge 0, y\ge 0)$, the field $[0,2]\times[0,2]$ can be regarded as the minimal envelop hyperbox. Any point in the MEH is not bound to be in the MVRL, but any point outside the MEH is bound to be outside the MVRL.

Definition 3: The maximal included hyperbox (MIH) and its opposite hyperbox. The MIH is wholly contained in the MVRL and its diagonal is in the superposition state with the diagonal of the MEH. We can attain the MIH through dividing the interval of the MEH's each dimension with the same proportion. We give the rigid mathematics definition of the MIH as follow. Set a coefficient $K\,(K\in[0,1])$ and let $\omega=[0,KV_1']\times[0,KV_2']\times\text{L}\times[0,KV_n']$ and $\omega\subseteq Pl$. $[0,K'V_1']\times[0,K'V_2']\times\text{L}\times[0,K'V_n']\not\subset Pl\ \forall K'>K,$. Furthermore, we define $\bar{\omega}$ as the opposite hybperbox of ω and the expression of $\bar{\omega}$ is described as $\bar{\omega}=(KV_1',\ V_1']\times(KV_2',\ V_2']\times\text{L}\times(K'V_n',\ V_n']$.

For the example mentioned above, $[0,2]\times[0,2]$ is the minimal envelop hyperbox (MEH). $[0,\sqrt{2}]\times[0,\sqrt{2}]$ is the maximal included hyperbox (MIH) and $(\sqrt{2},2]\times(\sqrt{2},2]$ is the opposite hyperbox with $K=\sqrt{2}/2$.

Theorem 1: Any point in the opposite hyperbox of the maximal included hyperbox does not meet the system requirement, i.e., is not in the MVRL.

Let $\omega=[0,KV_n']\times[0,KV_1']\times\text{L}\times[0,KV_2']$

Assume $\exists p\in\bar{\omega}=(KV_1',V_1']\times(KV_2',V']\times\text{L}\times(KV_n',V_n'],p=(p_1,p_2,\text{L},p_n)$ and $p\in Pl$

Let

$$Kp=\min\left(\frac{p_1-KV_1'}{V_1'},\frac{p_2-KV_2'}{V_2'},\text{L},\frac{p_n-KV_n'}{V_n'}\right)$$

($kp>0$ according to Definition 3)

$$\forall j,(Kp+K)V_j'\le\left(\frac{p_j-KV_j'}{V_j'}+K\right)V_j'=p_j$$

Let

$p_k=((Kp+K)V_1',(Kp+K)V_2',\text{L},(Kp+K)V_n')$ and $f_u((Kp+K)V_1',(Kp+K)V_2',\text{L},(Kp+K)V_n')\le f_u(p_1,p_2,\text{L}\ p_n)$ (according to the monotonic property of f_u).

In addition, $p \in Pl$, so $p_k \in Pl$. With the monotonic property of f_u and $\forall p' \in [0,(Kp+K)V_1'] \times [0,(Kp+K)V_2'] \times \text{L} \times [0,(Kp+K)V_n']$, $f_u(p_1', p_2', \text{L}, p_n') \le f_u(p_{k1}, p_{k2}, \text{L}, p_{kn})$.

So

$$[0,(Kp+K)V_1'] \times [0,(Kp+K)V_2'] \times \text{L} \times [0,(Kp+K)V_n'] \subseteq Pl$$

which contradicts the assumption that $\omega = [0, KV_1'] \times [0, KV_1'] \times \text{L} \times [0, KV_2']$, is the maximal included hyperbox since $Kp + K > K$.

The Algorithm

This algorithm recursively approximates MVRL with MIHs, and the algorithm steps are as follows.

Step 1 Determine the requirement u_k for Pl (MVRL), $f_u(Pl) \le u_k$

Step 2 Find the minimal envelop hyperbox (MEH):

Step 2.1 Set $k = 1$;

Step 2.2 All dimensions except the kth dimension are set to 0 (the minimal value). Due to the monotonic property of f_u, binary method is applied to get the upper bound of the kth dimension;

Step 2.3 If $k = n$, the cross product of interval [0, upper bound] of each dimension is the minimal envelop hyperbox (MEH), go to Step 3, else, set $k = k + 1$ and go to Step 2.2.

Step 3 Find the MIH:

Step 3.1 Judge whether the volume of MEH is smaller than the value E (which represents the exit condition). If true, make interpolation and exit. If the interval length of one certain dimension in the MEH is smaller than a certain value, this dimension need not be divided anymore and thus will be deleted;

Step 3.2 According to the monotonic property of f_u, it is obvious that when $K_1 \ge K_2$, $f_u(K_1V_1', K_1V_2', \text{L}, K_1V_n') \ge f_u(K_2V_1', K_2V_2', \text{L}, K_2V_n')$.

So, the binary method is applied to get the K (the dividing ratio for MIH) and the MIH;

Step 3.3 Each dimension interval of the MEH is divided into two parts. One part is inside the MIH and marked by "0", and the other is outside the MIH and marked by "1". So, there are altogether 2^n (n is the number of dimensions) interval combinations, each of which is denoted by a binary code containing n digits. The numerical value of the binary code is from 0 to 2^n-1. Code "0" whose binary digits are all 0 represents the MIH where any point meets the system requirement and will be saved. Code "1" whose binary digits are all 1 represents the opposite hyperbox of the MIH where any point does not meet the system requirement according to Theorem 1 and will be removed. Regard each of the rest combinations as a new MEH and recursively go to Step 3.

Figure 4 illustrates the algorithm process of acquiring the MVRL. Assume that the MVRL is the point set

$\{(x,y)|x^2+y^2\leq r^2;x\geq 0;\ y\geq 0\}$ and the rectangle marked by code "00" is the so-called MIH. Firstly, using the binary searching method along the diagonal of MEH, we can find the proper point in it and this point will finally locate the MIH. Secondly, get rid of the opposite hyperbox (marked by code "11") and regard each of the other hyperboxes as a MEH. Finally, acquire all the MIHs of which the MVRL is composed through repeating the process recursively.

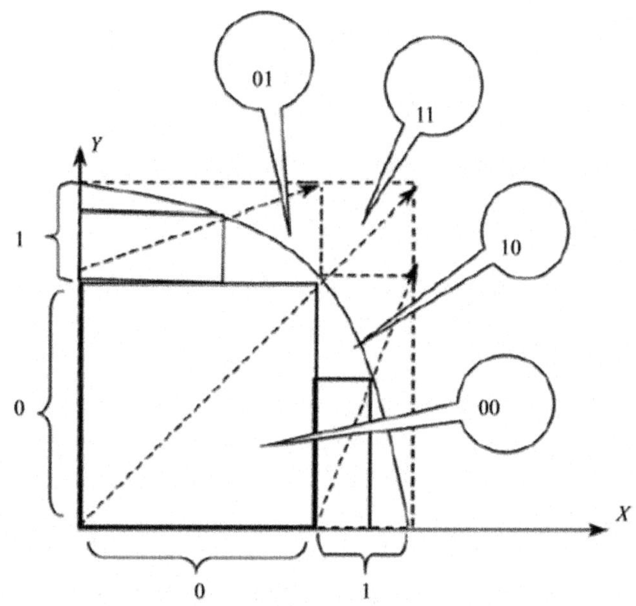

Figure 4: The process of acquiring the MVRL.

There exist monotonic realtions in MVS, thus we can get the special benefit of high efficiency as many famous algorithms which reduce calculation based on monotonicity (i.e. apriori algorithm in data mining [10], binary searching algorithm, branch and bound algorithm etc.)

This algorithm can be applied to solve many problems in which there exists the monotonicity. For example,it can be applied to efficiently draw the radar detecting range when the radar is jammed by jammer and efficiently seek the pareto locus in multi-object programming.

This algorithm is essentially a partition technique for some special figures. If the assumptions is met, it is superior to the classic triangulation method [11,12] by triangular or tetrahedron for the following reasons:

1) As the above mentioned, hyperbox is more suitable for many kinds of operations than triangular or tetrahedron.

2)It is effective and efficient to partition 3-dimensional or more than 3-dimensioal figure for which triangulation method is very difficult to be applied [12].

THE ALGORITHM FOR PARTITIONING OPERATION IN MVS OF DISCRETE DETERMINISTIC TYPE

An effective algorithm for partitioning in MVS of discrete type is proposed in this section. Based on the assumption that each dimension of MVS is divided into several segments, this algorithm can efficiently find out the combinations which meet the specified requirements. If each dimension of m-dimension MVS is divided into n segments, each of which is actually represented by the upper limit value of its own, there will be n^m combinations. According to the rule that segment with larger dimension value is tagged by lager mark, all the segments of each dimension are labeled with marks #1, #2, #3,···, and #n. The details of related algorithm are stated as follows.

Step 1 Specify a representative value (RV) for each combination and fill all the RVs into the combination table. RVs will be used to determine which combination is the next one to be judged whether meeting the requirement or not. Definition of RV is based on two other basic conceptions, i.e., the lower-cut set and the uppercut set. Explanations for these conceptions will be given with the following example. For example, if the combination #3#3#3#4 with m = 4 and n = 5 meet the specified requirement (i.e., $f_u(p_1,p_2,\mathrm{L}\ ,p_n)\le u_k, u_k\ge 0$ and f_u being monotonically increasing), there will be other 108 combinations $(3\times3\times3\times4=108)$ which can also meet the requirement. Every digit of each of 108 combinations (e.g., #1#1#1#1, #1#2#2#2, and #3#3#3#3.) is less than or equal to the corresponding digit of the combination #3#3#3#4, and all these 108 combinations constitute the so-called lower-cut set of combination #3#3#3#4. Similarly, if the above combination doesn't meet the requirement, there will be other 54 combinations $(3\times3\times3\times2=54)$ which cannot meet the requirement either. And all these 54 combinations (e.g., #4#4#4#4, #4#4#3#5, and #5#5#5#5.) constitute the socalled upper-cut set of combination #3#3#3#4. To determine an appropriate representative value for a certain combination, it may be helpful to take its lower-cut set and upper-cut set into account. So the calculation formula for RV can be stated as follows.

$$RV = NUPC \times INLOC/(INLOC + INUPC) + NLOC \times INUPC/(INLOC + INUPC).$$

In the above formula, $NLOC$, $NUPC$, $INLOC$, and $INUPC$ respectively denotes the present total element number of lower-cut set, the present total element number of upper-cut set, the initial total element number of lower-cut set, and the initial total element number of upper-cut set. For example, at the very beginning, the RV of combination #3#3#3#4 is $(108 \times 54/(108 + 54) + 54 \times 108/(108 + 54) = 72)$. When a certain combination meet the requirement, all the elements in its lower-cut set are cut. Otherwise, all the elements in its upper-cut set are cut. In this sense, RV reveals the calculation burden that may be alleviated by the corresponding combination.

Step 2 Scan the combination table, which is dimensioned by a heap structure to make the combination with the maximal RV found quickly, and select the combination with the maximal RV by using the greedy tactics. Firstly, assume combination #I1#I2#I3#I4 is the one with the maximal RV at present.

1) If the combination #I1#I2#I3#I4 meets the specified requirement, all the elements in its lower-cut set will be deleted from the combination table, which likely makes the lower-cut set element number of any other combination decreased. The method for adjusting the lower-cut set element number of any other combination (such as #J1#J2#J3#J4) is not so complicated. With setting #Ki =min(#Ii, #Ji)$^{(i=1,2,3,4)}$, the adjusted lower-cut set element number of combination #J1#J2#J3#J4 equals to subtracting the lower-cut set element number of combination #K1#K2#K3#K4 from the original lower-cut set element number of combination #J1#J2#J3#J4. If #K1#K2#K3#K4 is not in the combination table, the original lower-cut set element number of combination #J1#J2#J3#J4 will remain unchanged.
2) If the combination #I1#I2#I3#I4 does not meet the specified requirement, all the elements in its upper-cut set will be deleted from the combination table, which likely makes the upper-cut set element number of any other combination decreased. The method for adjusting the upper-cut set element number of any other combination (such as #J1#J2#J3#J4) is not so complicated. After setting #Ki = min(#Ii, #Ji) $^{(i=1,2,3,4)}$, the adjusted upper-cut set element number of combination #J1#J2#J3#J4 equals to subtracting the upper-cut set element number of combination #K1#K2#K3#K4 from the original upper-cut set element number of combination #J1#J2#J3#J4. If #K1#K2#K3#K4 is not in the combination table, the original upper-cut set element number of combination #J1#J2#J3#J4 will remain unchanged.

Step 3 Adjust the combination table. If the selected combination meets the specified requirement, this combination and all the elements in its lower-cut set will be deleted from the combination table, or else this combination and all the elements in its upper-cut set will be deleted from the combination table. For example, if combination #3#3#3#4 meets the specified requirement, all the elements in its lower-cut set such as #1#1#1#1, #1#2#2#2, #3#3#3#3, etc. in the combination table would be deleted.

Step 4 Go to step 2 until the combination table is empty.

To illustrate the above algorithm in detail, it is necessary to discuss an example. In the following example, a discrete MVS with four dimensions (i.e., x_1, x_2, x_3 and x_4) is assumed and each dimension is divided into four segments whose upper limit values are respectively 2(#1, No.1), 3(#2, No.2), 4(#3, No.3), and 5(#4, No.4). And the requirement condition is specified as $x_1^2+x_2^2+x_3^2+x_4^2\leq 50$.

Now the partitioning operation can be carried out under the guidance of the above algorithm.

Step 1 Determine the corresponding RVs for all the combinations, the number of which is 256$(4^4=256)$ due to $m=4$ and $n=4$, and fill all the RVs into the combination table (shown in Figure 5).

Step 2 Single out the combination with the maximal RV from combination table. At the very beginning, combination #3#3#2#2 is found to be the combination with the maximal RV$(\mathrm{RV}=36)$.

Step 3 Adjust the combination table. For example, since combination #3#3#2#2 meets the specified requirement, on the one hand, this combination and all the elements in its lower-cut set are to be deleted from the combination table; on the other hand, the lower-cut set element number of any remaining combination in the combination table may also need to be modified. Now take the combination #4#1#1#1 as an example. Combination

#K1#K2#K3#K4 = #3#1#1#1 with #I1#I2#I3#I4 = #3#3#2#2 and #J1#J2#J3#J4 = #4#1#1#1 can be easily gotten by using the method mentioned above. Because combination #3#1#1#1 is found to be still in the combination table, the lower-cut set element number and RV of combination #4#1#1#1 must be modified. So the adjusted lower-cut set element number of #4#1#1#1 is equal to its original lower-cut set element number minus the lower-cut set element number of combination #3#1#1#1 (i.e., $4-3=1$). Based on the adjusted lower-cut set element number of #4#1#1#1, its adjusted RV can also be acquired. The adjusted lower-cut set element numbers and adjusted RVs of other combinations are shown in Table1

Combination No	Total element number of lower-cut set	Total element number of upper-cut set	RV
#1#1#1#1	1	256	1.99
#2##1#1#1	2	192	3.96
#3# 1#1#1	3	128	5.86
……			
#3#3#2#2	36	36	36
……			
#4#1#1#1	4'	64	7.53
#2#4#4#4	128	3	5.86
#3#4#4#4	192	2	3.96
#4#4#4#4	256	1	1.99

Find the combination #3#3#2#2 with the maximal RV in the combination table and then delete all the elements in its lower-cut set from the combination table.

The lower-cut set of combination #3#3#2#2
#1#1#1#1
#2#1#1#1
#3#1#1#1
#3#2#1#1
#3#2#2#1
#3#3#1#1
……

Figure 5: Combination table.

Step 4 Go to step 2 until the combination table is empty. In this example, all the combinations meeting the specified requirement such as #1#2#2#4, #2#1#2#4,···, #3#3#2#2, etc., are acquired after 67 iterations, which means that only about one quarter of all the combinations are calculated and the algorithm elaborated in this section is of high efficiency.

THE ALGORITHM FOR SYNTHESIZING OPERATION IN MVS OF CONTINUOUS TYPE

For a practical problem, there are usually several different kinds of requirement targets, each of which obviously corresponds to a different MVRL. Therefore, intersection operation of multiple MVRLs is of great significance to the solution of one practical problem. In this section, two algorithms for one kind of synthesizing operations (i.e., intersection operation of multiple MVRLs) are presented.

Algorithm Based on Direct Method

The shape of a MVRL approached by many hyperboxes is usually irregular. The intersection of multiple MVRLs is approximately to be the intersection of the hyperboxes included in the MVRLs. Therefore, the former can be acquired through finding out the latter. The method of seeking the total intersection of all the hyperboxes is very simple. However, its time cost is large. For example, if there are two MVRLs whose numbers of hyperboxes are respectively M and N, $M \times N$ intersection operations have to be performed to get the total intersection of all the hyperboxes. And the time complexity is $O(n^2)$, where n the hyperbox number of one MVRL is.

Actually, all the MEHs and MIHs acquired in the process of producing a MVRL can be naturally organized into an including relation tree (heap) shown in Figure 6.

Table 1: Combination table

Cobination No.	Total lower-cut set Element number	Total upper-cut set element number	RV
#4#1#1#1	1*(have been decreased)	64	4.7
#4#2#1#1	2	48	8.57
#4#3#1#1	3	16	10.9
......			
#4#4#4#4	220	2	1.85

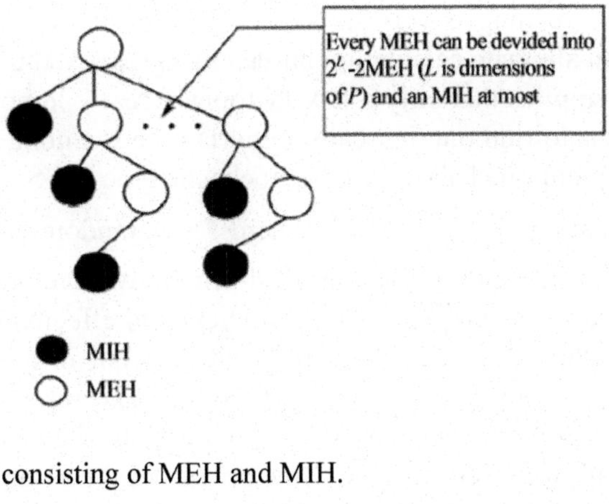

Figure 6: The tree consisting of MEH and MIH.

Algorithm Based on Approaching Method

If all the requirements have the same constraint direction on any system dimension, i.e., achievement degree of all the requirement indices becoming larger or smaller with any system dimension value increasing or decreasing, approaching method can be used to seek the intersection of multiple MVRLs. The algorithm based on approaching method is stated as follows.

Step 1 Get the intersection of the MEHs of all the MVRLs, i.e., the MEH of the intersection of all the MVLRs.

Step 2 Seek all the MIHs, which approximately compose the intersection of all MVRLs, with the related algorithm mentioned above.

The above two steps are illustrated in Figure 7.

If this tree is saved in the memory, the time cost of the direct method may be largely reduced. The reason is that if the hyperbox of one MVRL has null intersection with one certain MEH of another MVRL, it will have no common intersection with all the MIHs included in this MEH, and so, the related intersection operations can be omitted.

Besides intersection operation, which is a relatively simple and typical synthesizing operation, there are still some other complicated synthesizing operations such as union operation, union-intersection mixed operation (i.e., carrying out intersection operation on some dimensions and union operation on the others). And all of these synthesizing operations are very useful in solving practical problems and should be studied thoroughly in the future.

THE ALGORITHM FOR SCREENING OPERATION IN MVS OF STOCHASTIC TYPE

Dimensions screening is very important for all kinds of operations in MVS because it can alleviate calculation burden to a great extent. Pareto principle or 20 - 80 rule implies that only a few factors are really important. The purpose of screening is to find out the really important ones among all the dimensions of MVS. As mentioned above, in the stochastic type MVS, the mapping f_u will be of statistical type, i.e., $u = f_u(P) + \varepsilon$ where ε is a random value and $E(\varepsilon) = 0$.

Sequential bifurcation (SB) method originally developed by Bettonvil in his doctoral dissertation [4] is a very efficient and effective one among all the screening methods and will be introduced in detail in this section. And the summarization of Bettonvil's doctoral dissertation is presented by reference[5]. Assuming a specific metamodel, an approximation of the monotonic mapping f_u, is necessary for a screening. Characteristics of the assumed metamodel for SB in this part include two points.

1) The metamodel of f_u is of monotonic type, which is obviously met in MVS.
2) The metamodel is a first-order polynomial with $u = \beta_0 + \beta_1 p_1 + \beta_2 p_2 + \cdots \beta_n p_n + \varepsilon$ and $E(\varepsilon) = 0$.

The dimensions $p_j\ (j = 1, \cdots, n)$ of MVS are standardized such that they are either −1 or +1 which represent the maximal or minimal value for p_j. This scaling implies that the dimensions may be ranked (sorted) by their main effects; i.e., the most important dimension is the one with the largest absolute value of its first-order effect or main effect; the least important dimension is the one with the effect closest to zero.

To explain some mathematical details of SB, the following notations are used.

$w(j);r$: observed output with the dimension 1 through j set to their high levels and the remaining factors set to their low levels, in replication r;

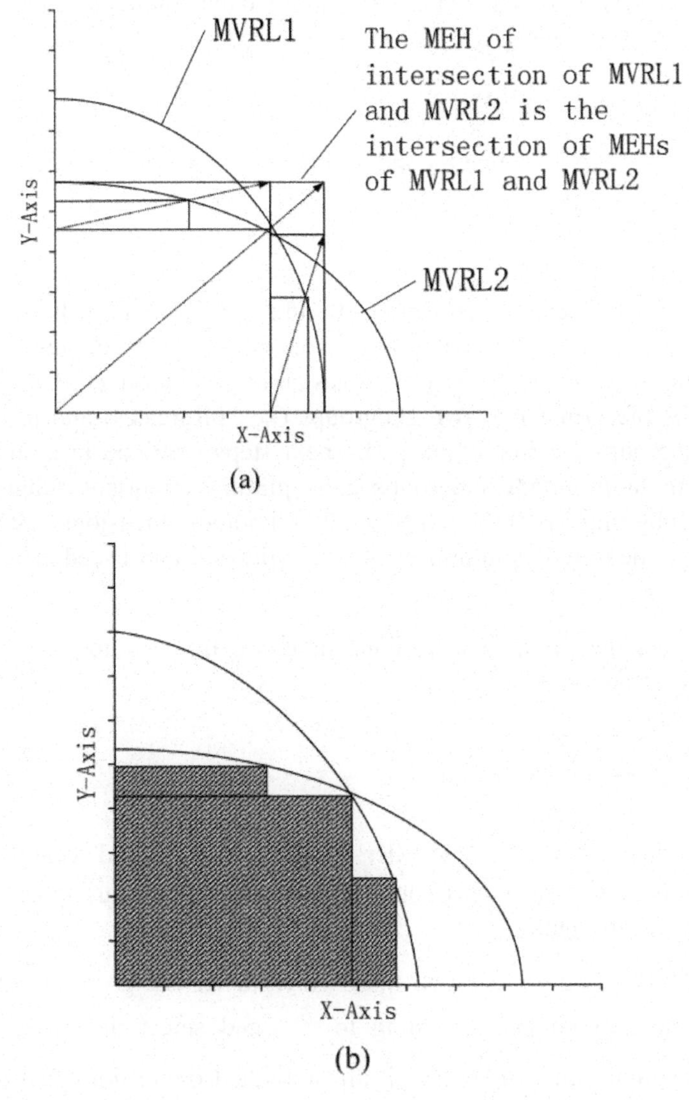

Figure 7: The approaching method to acquire the intersection of MVRL1 and MVRL2.

$\beta_{j'-j}$: sum of main effects of dimension j' through j; so $\beta_{j'-j} = \sum_{h=j'}^{j} \beta_h$

A simple estimate of this group effect based on replication r is $\hat{\beta}_{j'-j;r} = \frac{w_{(j);r} - w_{(j'-1);r}}{2}$

SB starts with calculating the two most extreme scenarios: in scenario 1, all k dimensions are set at their low levels so $p_j = -1$; in scenario 2, all these factors are high so $p_j = +1 \left(j = 1, \mathrm{L}, k \right)$. If the above metamodel is valid, then $E\left(w_{(0)}\right) = \beta_0 - \beta_1 - \mathrm{L} - \beta_k$ and $E\left(w_{(k)}\right) = \beta_0 + \beta_1 + \mathrm{L} + \beta_k$ so $E\left(w_{(k)}\right) - E\left(w_{(0)}\right) = 2(\beta_1 + \mathrm{L} + \beta_k)$.

Likewise it follows that the individual main effect is estimated:

$$\hat{\beta}_{j'-j;r} = \frac{w_{(j);r} - w_{(j-1);r}}{2}$$

The SB procedure is sequential. Its first step places all dimensions into a single group, and tests whether or not that group of dimensions has an important effect. If the group indeed has an important effect, then the second step splits the group into two subgroups (i.e., bifurcates) and tests each of these subgroups for importance. The next steps continue in a similar way, discarding unimportant subgroups and splitting important subgroups into smaller subgroups. In the Final step, all individual dimensions that are not in subgroups identified as unimportant, are estimated and tested. The steps are showed below:

1) Select the group whose dimensions are from j' to j and estimate expected value of the

$$\hat{\beta}_{j'-j} = \frac{1}{r} \sum_{k=1}^{r} \hat{\beta}_{j'-j;k};$$

2) If this group's $\hat{\beta}_{j'-j}$ smaller than k (a threshold value), then all individual dimensions(from j' to j) that are in this group are identified as unimportant;

3) If this group's $\hat{\beta}_{j'-j}$ larger than k, then insert this group into the group table ascendingly according to $\hat{\beta}_{j'-j}$ and select the group with the maximal value of $\hat{\beta}_{j'-j}$, until all individual dimensions that are not in subgroups identified as unimportant are estimated and tested.

This section introduces a simple algorithm (i.e., sequential bifurcation) for screening operation in MVS. SB method based on some assumptions is effective and efficient in the solving of many practical problems(i.e., supply chain analysis [6), And some investigator, such as the authors of [7-9] have successfully extended the simple SB.

THE SAMPLING OPERATION IN MVS OF STOCHASTIC TYPE

Sampling operation is important for experiments in MVS. The purpose of this operation is to effectively and efficiently select the points used in all kinds of experiments. Latin hypercube sampling (LHS) is very fit for MVS. The algorithm for acquiring n sampling points in the d-dimension hypercube $[0,1]^d$ with LHS method is discussed in this section. Coordinate value interval $[0,1]$ of each dimension is equally divided into n segments and the ith segment $\left(\frac{i-1}{n},\frac{i}{n}\right]$ is marked by its segment number i $(i=1,L,n)$. A random permutation of n segment numbers of the kth dimension is denoted as π_{1k},L,π_{nk} $(k=1,L,d)$. And d mutual independent random permutations constitute a $n\times d$ random matrix $\pi\left(\pi=(\pi_{ij})\right)$. Such n sampling points as $P_i=(P_{i1},L,P_{id})$ $(i=1,L,n)$ with $P_{ij}=\left(\pi_{ij}-\frac{1}{2}+u_{ij}\right)/n$ $(i\in\{1,L,n\};j\in\{1,L,d\})$ where u_{ij} is random variable subject to uniform distribution on $[0,1]$ and not only mutual independent but also independent of π construct a LHS sample. In many aspects, a LHS sample is superior to some other kinds of samples such as the independently and identically distributed (IID) sample gotten with simple random sampling (SRS) method.

The relative advantages of LHS method can be revealed with an example. The estimation value $\left(\bar{U}=\frac{1}{n}\sum_{i=1}^{n}f_u(P_i)\right)$ of the mathematical expectation $E(f_u(P))$ of function $U=f_u(P)$ will be different with different sample. And the estimation values based on IID sample, LHS sample, and the sample acquired with stratified sampling (SS) method are respectively denoted as $\bar{U}_R$, $\bar{U}_L$, and $\bar{U}_S$. The reference [3] has proved the unbiasedness of these estimations and also given the variances of $\bar{U}_R$, $\bar{U}_L$, and $\bar{U}_S$.

$$\begin{cases} Var(\bar{U}_R) = \frac{1}{n} Var(U); \\ Var(\bar{U}_S) = Var(\bar{U}_R) - \frac{1}{n^2}\sum_{i=1}^{n}(u_i - u)^2; \\ Var(\bar{U}_L) = Var(\bar{U}_R) + \frac{n-1}{n} \cdot \frac{1}{n^k (n-1)^k}\sum_R (u_i - u)(u_j - u). \end{cases}$$

In the above expressions, $u = E(f_u(P))$, $u_i = E(U|P \in S_i)$ for LHS sample, and $u_i = E(U|P \in gridi)$ for SS sample (referring to [3] for other details). Since $f_u(p_1, \ldots, p_d)$ will monotonically increase or decrease with each of d indices changing monotonically while the others maintaining to be constants in MVS, such conclusion as $\sum_R (u_i - u)(u_j - u) \le 0$ can be gotten. Actually, the above conclusion, with which another conclusion (i.e., $Var(\bar{U}_L) \le Var(\bar{U}_R)$) can be deduced, has already been proved by reference [3]. Thus it can be seen that LHS is a sampling method fit for experiments in MVS.

REFERENCES

1. J. W. Hu, et al., "MIS Method (Monotonic Indices Space) and Its Application in Capability Indicators Effectiveness Analysis of a Notional Anti-Stealth Information System," IEEE Transactions on Systems, Man and Cybernetics PART A, Vol. 39, No. 2, 2009, pp. 407-417.
2. V. Bouthonnier and A. H. Levis, "System Effectiveness Analysis of C3 Systems," IEEE Transactions on Systems, Man and Cybernetics, Vol. SMC-14, No.1, 1984, pp. 48-54.http://dx.doi.org/10.1109/TSMC.1984.6313268
3. M. D. McKay, et al., "A Comparision of Three Methods for Selecting Values of Input Variablea in Analysis of Output from a Computer Codes," Tehchnometrics, Vol. 21, No. 2, 1979, pp. 239-245.
4. B. Bettonvil, "Detection of Important Factors by Sequential Bifurcation," Ph.D. Dissertation, Tilburg University Press, Tilburg, 1996.
5. B. Bettonvil and J. P. C. Kleijnen, "Searching for Important Factors in Simulation Models with Many Factors: Sequential Bifurcation," European Journal of Operational Research, Vol. 96, No. 1, 1997, pp. 180-194. http://dx.doi.org/10.1016/S0377-2217(96)00156-7
6. J. P. C. Kleijnen, B. Bettonvil and F. Persson, "Screening for the Important Factors in Large Discrete-Event Simulation: Sequential Bifurcation and

Its Applications," In: A. Dean and S. Lewis, Eds., Screening: Methods for Experimentation in Industry, Drug Discovery, and Genetics, Springer-Verlag, New York, 2006, pp. 287-307. http://dx.doi.org/10.1007/0-387-28014-6_13

7. H. Wan, B. E. Ankenman and B. L. Nelson, "Controlled Sequential Bifurcation: A New Factor-Screening Method for DiscreteEvent Simulation," Operations Research, Vol. 54, No. 4, 2006, pp. 743-755. http://dx.doi.org/10.1287/opre.1060.0311

8. R. C. H. Cheng, "Searching for Important Factors: Sequential Bifurcation under Uncertainty," Proceedings of the 1997 Winter Simulation Conference, 1997, pp. 275-280.

9. B. E. Ankenman, R. C. H. Cheng and S. M. Lewis, "A Polytope Method for Estimating Factor Main Effects Efficiently," Proceedings of the 2006 Winter Simulation Conference, Monterey, 3-6 December 2006. http://dx.doi.org/10.1109/WSC.2006.323104

10. R. Agrawal and R. Srikant, "Fast Algorithms for Mining Association Rules," Proceedings of 1994 International Conference Very Large Data Bases (VLDB'94), Santiago, September 1994.

11. B. Chazelle and L. Palios, "Triangulating a Non-Convex Polytope," Discrete & Computational Geometry, Vol. 5, No. 1, 1990, pp. 505-526. http://dx.doi.org/10.1007/BF02187807

12. J. Ruppert and R. Seidel, "On the Difficulty Triangulating Three-Dimensional Non-Vex Polyhedra," Discrete & Computational Geometry, Vol. 7, 1992, pp. 227-253.

Chapter 5

MONOTONIC VECTOR SPACE MODEL(II): APPLICATION IN SYSTEM CAPABILITY INDICATOR ANALYSIS AND DESIGN

Jianwen Hu

National University of Defense Technology, Changsha, China

ABSTRACT

The application of some MVS operations (each dimension of which corresponds to a capability indicator) in sys- tem capability analysis and design, including capability indicator requirement analysis, effectiveness analysis, sensibility analysis, fuzzy analysis, stability analysis, capability optimization design, etc., is discussed in the second paper of this series of papers. And some MVS-based models and algorithms for capability analysis and design are put forward. Finally, an example of capability analysis and optimization design is given for explaining the usages of related models and algorithms.

INTRODUCTION

System capability indicator analysis and design are very important and also difficult for system development and system engineering. Many practical methods have been proposed and will be introduced in the following.

The Representative System Capability Indicator Analysis Methods

- Index method. It is a traditional method for weapon system capability indicator analysis. This method is suit- able for synthesis and easy to understood. However, the index formula and weight influenced by analyzer are difficult to be determined. And the analysis result of index method is to some extent ambiguous.
- ADC model [1]. Based on the principle of division and transformation of system state, this method proposed by WEISIC is able to give clear quantitative analysis result. However, the calculation overload will

exponen- tially grow with the increasing of system dimension number. Furthermore, other models and algorithms are required to get the capability vector C of ADC model.

- Weight-based multi-attributes analysis method [2,3]. This type of method such as AHP [4], ELECTRE, LINAMP, etc., determines weight of each indicator firstly. Then, all indicators are synthesized according to each of their weights in a linear way. Obviously, they are regardless of the mutual influence among indicators and so fail to reveal the nonlinearity of complex system. And furthermore their analysis results are also unclear.
- SEA method [5]. This method compares system capabilities with mission requirements in a common attribute space and emphasizes the whole characteristic of complex system. This comparison leads to the evaluation of partial measures of effectiveness that are then combined to yield a global measure. But system mission requirement loci used by this method are difficult to acquire.

System Capability Indicator Design Method

- Linera programming methods. This kind of method is traditional and neat. Many complex problems (especialy with nolineartity and uncertainty) have to be over trivialized so as to be solved by these neat methods. Therefore, these methods are not directly effective for many complex problems.
- Capability-based Planning (CPB) [6]. CPB method emphasizes the flexibility, adaptiveness and robustness of capability and usually adopts exploratory analysis strategy. Compared with programming method, CPB, as a flexible and non-neat method, is more suitable for complex problems.

Monotonic vector space (MVS) model is used for system capability indicator analysis and design in this paper. Since each dimension of MVS corresponds to one capability indicator, the MVS, therefore, is also called Monotonic Indicator Space (MIS) here. And likewise there exist some monotonic mappings between capability in- dicators and requirement measures in MIS.

Assumptions Adopted in This Paper

A monotonic relationship between the system capability indicators and the system requirement measures is assumed in this paper. It must be noted that system capability indicators are different from system parameters. For example, the velocity of a certain aircraft is not a capability indicator, but the maximal (or minimal) velocity of the aircraft is a system capability indicator. This assumption is usually in accordance with the practical situation in many cases. For example, in reference [5], there exists a monotonic relationship between the capability indi- cators (including reliability indicator R, survivability indicator S, delay time indicator T, and link capacity indicator K) and the requirement measure, i.e. the ratio of kill to loss. In reference [7], there also exists a monotonic relationship between the capability indicators (including link reliability indicator and delay time indicator) and the requirement measure (the kill probability). Actually, similar monotonic relationship between the capability indicators and the requirement measure can be found in many practical cases introduced by references [8-12].

In addition, the weak capability is assumed to be a subset of the stronger one. Thus, if a system with weak capability meets some certain mission requirements, the one with stronger capability will be certain to meet these mission requirements. For example, if an aircraft with the maximal cruise velocity of 1800 km/h can ar- rive at a certain location in one hour (a requirement measure), the one with maximal cruise velocity of 3000 km/h will definitely be able to arrive at the same location in one hour as long as the other capability indicators of these two aircrafts are equal.

The contents of system capability indicator analysis and design based on MIS include indicator requirement analysis, effectiveness analysis, sensitivity analysis, fuzzy analysis, relativity analysis, finance-based optimiza- tion design and time-based optimization design. The mutual relationships among these contents are shown in Figure 1. Obviously, system indicator analysis is the start point of system capability indicator analysis and de- sign. And in the following sections, the related models and algorithms for system capability indicator analysis and design will be dealt with in detail.

MIS-BASED SYSTEM CAPABILITY INDICATOR ANALYSIS

System Capability Indicator Requirement Analysis

The partitioning operation in MVS introduced by the first part of this series paper is used to produce the system capability requirement locus. For example, the dimensions of a monotonic indicator space P for a ground-to-air missile system may include the searching radar detecting capability indicator dimension K_1, delay time indicator dimension T, and tracking radar detecting capability indicator dimension K_2. So, the equation

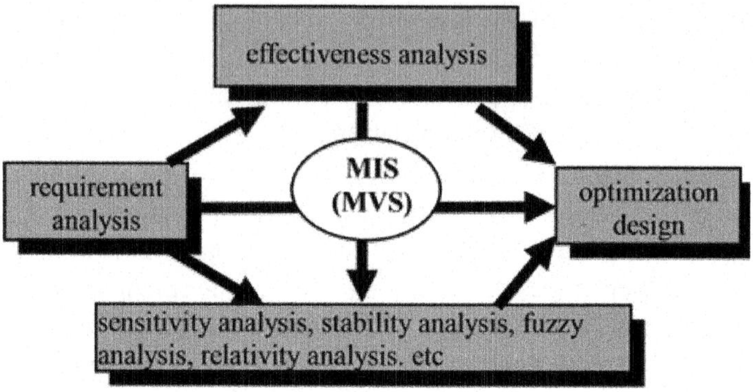

Figure 1: MVS-based system capability indicators analysis and design.

$u = f(K_1, T, K_2)$ describes the relationship between these three system capability indicators (i.e., K_1, T, and K_2) and one requirement measure u (the kill probability). The function f may adopt several different forms, each of which corresponds to a different context. For example, f_1 and f_2 correspond to the different contexts in which the enemy stealth aircraft is respectively type A and type B. Function f is obviously monotonic, that is to say, the value of u (the kill probability) will monotonically increase or decrease with one of three system capability indicators changing monotonically while the others maintaining to be constants. Because the mono- tonic relationship between capability indicators (K_1 and K_2) and the requirement measure (the kill probability) is different from that between capability indicator (T) and the kill probability, so, K_1 and K_2 are called monotonic increasing indicators and T called monotonic decreasing indicator. Every dimension in MIS can be continuous or discrete type. System capability indicators are usually of continuous type as all the above mentioned ones. For discrete-type indicators, if its range is large enough, it still can be considered as continuous type just like many integer programming problems can be solved by general linear programming model. In this paper, all the dimensions of MIS are assumed to be continuous. The purpose of system capability indicator requirement analysis is to find out the MVRL or MIRL (monotonic indicator requirement locus) which can then be acquired with the algorithm proposed by the first part of this series papers. Figure 2 shows the MIRL for the above-mentioned ground-to-air missile system. There are 3 MIS dimensions, each of which corresponds to a capability indicator. The requirement measure u (the kill probability for enemy invading stealth aircraft of type A and type B) is assumed as $u \geq 0.7$. The simulation calculation realizes the mapping from capability indicators to requirement measure. The MIRLs are produced by the partitioning algorithm at first and then synthesizing operation is used to produce the intersection of these multiple MIRLs which correspond to multiple requirements. According to the algorithm for producing MIRL, the algorithm output is a heap tree consisting of MIHs and MEHs. The MIHs compose the MIRL.

Effectiveness Analysis for the Indicators of Stochastic Type

Due to the uncertainty of the system capability indicators, it is usually difficult to represent them precisely. So the probability distribution function is applied to represent the system capability indicators. We represent a system capability indicator with such equation as $p_i = p_{ie} + p_{is}$, where p_{ie} and p_{is} respectively denote the expected value and a stochastic variable. Furthermore, a union probability density function f(p) where p is the vector in the MIS is used to describe the system capability indicators. The system effectiveness is denoted as E and its expression is $E = \int_{Pl} f(p)\mathrm{d}p$, where Pl is the MIRL. It can be easily known that the range of E is [0,1] . As the system effectiveness, E reflects the extent to which system meets the requirements. If the MIRL is approximated by the hyperboxes, i.e. $Pl \approx \bigcup_{j=1}^{N} H_j$, where H_j is the jth hyperbox and N is the number of hyperboxes, the expression of E will change into $E \approx \sum_{j=1}^{N} \int_{H_j} f(p)\mathrm{d}p$.

Sensitivity and Relativity Analysis for the Indicators of Stochastic Type

The purpose of getting the indicator's sensitivity includes: 1) acquire the weightiness of indicator and determine which indicator is critical; 2) more carefully deal with the uncertain and sensitive context or system capability indicators; and 3) optimize system.

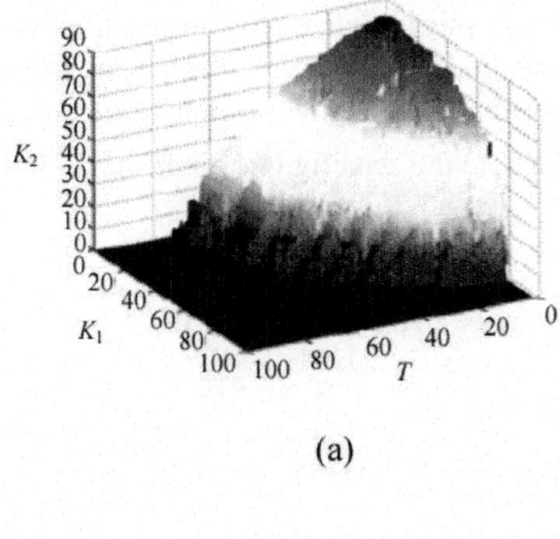

(a)

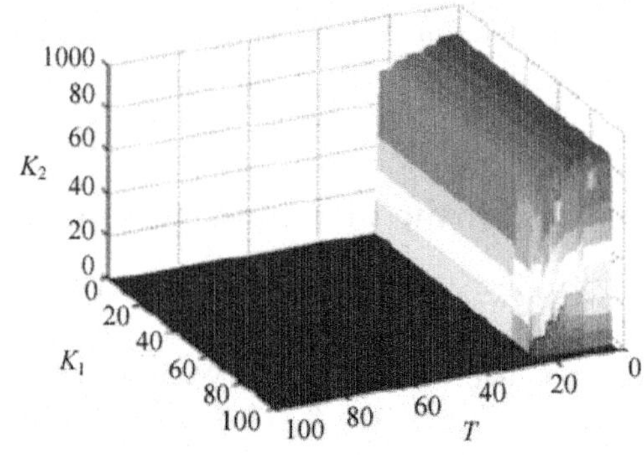

(b)

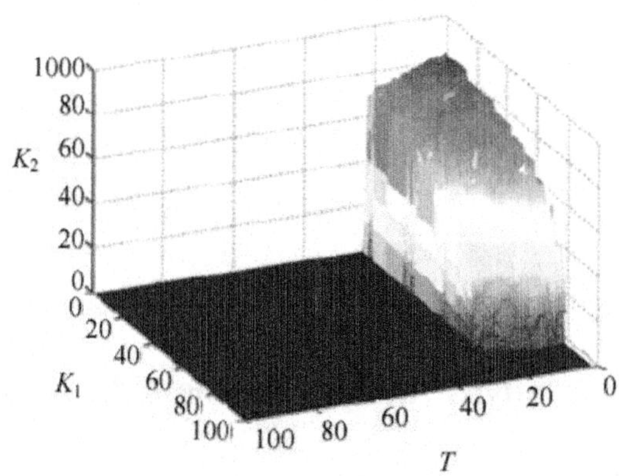

(c)

Figure 2: (a) MIRL1 for the stealth aircraft of type A; (b) MIRL2 for the type B; (c) the intersection of MIRL1 and MIRL2 acquired by the approaching method.

The sensitivity analysis is realized by the perturbation analysis. This paper uses the analytic method to acquire the indicator's sensitivity. In addition, every indicator is normalized within a common range for simplicity.

Definition 1: Denote the sensitivity of indicator p_i as S_i and its definition is

$$S_i = \frac{\partial E}{\partial e_i} = \frac{\partial\left(\int_{PI} f(p,e)\mathrm{d}p\right)}{\partial e_i} \approx \frac{\partial\left(\sum_{j=1}^{N}\int_{H_j} f(p,e)\mathrm{d}p\right)}{\partial e_i},$$

where $pl \approx \bigcup_{j=1}^{N} H_j$ and e_i is the expected value of the system capability indicator p_i (stochastic variable) and also the perturbation variable.

Definition 2: Denote the relativity between indicator p_i and p_j as R_{ij}.

$$\text{If } i = j, R_{ij} = \frac{\partial^2 E}{\partial e_i^2};$$

$$\text{If } i \neq j,\ R_{ij} = \frac{\partial^2 E}{\partial e_i \partial e_j};$$

Definition 3: Possibility measure [13]. θ is assumed as a non-empty set and P(θ) as the power set of θ . If Pos meets three axioms, i.e. $Pos(\theta) = 1$, $Pos(\Phi) = 0$, and $Pos\left\{\underset{i}{\cup} A_i\right\} = \sup_i Pos\{A_i\}(A_i \in P(\theta))$, it is called possibility measure

Definition 4: System effectiveness based on the fuzzy indicators and possibility measure E_{fp}. The expression of E_{fp} is $E_{fp} = Pos(p \in Pl)$, where p is multi-dimensional fuzzy indicator and Pl is the requirement locus. The value of E_{fp} can be acquired by fuzzy simulation method which is detailed as follows.

Step 1 Set $L = a$ and a is a very small estimate of L.

Step 2 According to subordinating fuction of p_i, uniformly produce $u_i\ (i = 1, 2, \cdots, n)$ from a-cut set and let $u = (u_1, u_2, \cdots, u_n)$.

Step 3 Set $u = u_1(p_1) \wedge u_2(p_2) \wedge \cdots \wedge u_n(p_n)$.

Step 4 If $p \in Pl$ and $L < u$, set $L = u$.

Step 5 Repeat step 2 to 4 N times.

Step 6 If $E_{fp} = L$, return E_{fp}.

Credibility, which reveals the possibility and impossibility simultaneously, is a novel mode describing fuzzy relation. So, as a measure, credibility is more reasonable than possibility. The definition of credibility measure is stated as follows.

Definition 5: Credibility measure [13]. With the assumption that $(\theta, P(\theta), Pos)$ is a possibility space and A is the element of power set P(θ), credibility measure Cr { A } is defined as

$$Cr\{A\} = \frac{1}{2}\left(Pos\{A\} + \left(1 - Pos\{A^c\}\right)\right).$$

Definition 6: System effectiveness based on the fuzzy indicators and credibility measure E_{fc}. The expression of E_{fc} can be stated as $E_{fc} = Cr(p \in Pl) = \frac{1}{2}\left(Pos\{p \in Pl\} + \left(1 - Pos\{p \notin Pl\}\right)\right)$, where p is multidimensional fuzzy indicator and Pl is the requirement locus. Similarly, it can be acquired by fuzzy simulation method.

Step 1 Uniformly produce (k) p from p and let let $v_k = Pos\{p^{(k)}\}(k=1,2,\cdots,N)$ if $Pos\{p^{(k)}\} \geq \varepsilon$, where ε is a very small number.

Step 2 Let $E_{fc} = \frac{1}{2}\left(\max_{1\leq k\leq N}\{v_k | p^{(k)} \in Pl\} + \min_{1\leq k\leq N}\{1-v_k | p^{(k)} \notin Pl\}\right)$ and then return E_{fc}. For the above two algorithms, it is important to judge whether the p is within Pl or not. And based on the tree structure output of partitioning algorithms introduced in the first one of this series papers, it can be quickly judged whether the p is within Pl or not. The judging procedure is detailed as follows. Firstly, judge whether the p is within the top MEH (root of the tree) or not. If it is not, it is not within Pl either and the judging procedure ends. If p is within the MEH, judge whether p is within the next-level MIH or not. If p is within the MIH, it is also within Pl and the judging procedure ends. If it is not, judge whether p is within one certain MEH of this level or not. If p is contained by one MEH, continue the recursive searching procedure from this MEH. If searching reaches the bottom of the tree and no one MIH containing p is found, it means that this p is not within Pl.

Effectiveness Analysis Based on Fuzzy MIRL

Definition 7: Fuzzy monotonic indicator requirement locus (FMIRL) $fpl(\lambda)$. $\tilde{R}$ is a fuzzy set on a certain domain; $\tilde{R}_\lambda$ is the l-cut set of $\tilde{R}$; and $fpl(\lambda)$ is the effective range in P (MIS) which meets the requirement $\tilde{R}_\lambda$. These concepts are now illustrated by an example. For an air defense system, the mission requirement measure (fuzzy set $\tilde{R}$) is killing the enemy aircraft. The domain is the kill probability. Set a l to determine the $\tilde{R}_\lambda$. $\tilde{R}_\lambda$ is a requirement measure of kill probability that must be larger than a certain value.

The fuzzy system capability indicator (assumed to be stochastic) effectiveness analysis model is introduced as follows. $fpl(\lambda)$ can be determined according to l. If $\lambda_1 \leq \lambda_2$, we can get the corresponding conclusion, i.e., $\tilde{R}_{\lambda_2} \subseteq \tilde{R}_{\lambda_1}$ and $fpl(\lambda_2) \subseteq fpl(\lambda_1)$ according to the monotonic assumption.

$E_f(\text{Fuzzy})\int_{pl} f(p)\mathrm{d}p = \bigvee_{0\leq\lambda\leq 1}\left(\lambda \wedge \int_{fpl(\lambda)} f(p)\mathrm{d}p\right)$ and; $\int_{fpl(\lambda_1)} f(p)\mathrm{d}p \geq \int_{fpl(\lambda_2)} f(p)\mathrm{d}p$ when $\lambda_1 \leq \lambda_2$. So, $E_f = \lambda^* = \int_{fpl(\lambda^*)} f(p)\mathrm{d}p$. And E_f can be solved rapidly by the binary method. The fuzzy effectiveness analysis model is not only suitable for the fuzzy requirement but also able to decrease misclassification rate of the boundary points especially when the system capability indicator values are distributed near the boundary of MIRL.

MIS-BASED SYSTEM CAPABILITY INDICATOR OPTIMIZATION DESIGN

Budget-Oriented Optimization Design

Budget-oriented capability indicator optimization design will be discussed in this section. In order to make a system meet a certain requirement, there are usually many development "paths" with different cost for system capability indicators. Figure 3 gives three linking paths (i.e., OA, OB and OC) between point O representing the initial system capability and the requirement locus. And point O is assumed to be out of the requirement locus. Certainly, besides these three paths, there exist countless other linking paths between point O and requirement locus. The purpose of capability indicator optimization design is to find out the path with the minimal budget or least time cost. In this paper, the optimization model is used to get the optimized expectation values of the statistical capability indicators according to practical problems. For simplicity, all the capability indicators are assumed to be mutually independent and have fixed variances (i.e., every capability indicator can be developed independently). And the optimization design model is:

$$\mathrm{Min}\left(C\left(P_0,P_e\right)\right)$$
$$\mathrm{s.t}: E \geq K, E = \int_{Pl} f\left(p,P_e\right)\mathrm{d}p.$$

In this model: $f\left(p,P_e\right)$ is union probability density function and P_e, as the optimization variable, represents the expectation value of $f\left(p,P_e\right)$; pl, E, and K respectively represent the MIRL (monotonic indicators requirement locus), the effectiveness value, and the capability indicator effectiveness requirement threshold; $C\left(P_0,P_e\right)$ represents the budget-cost function of capability indicator varying from P_0 to P_e.

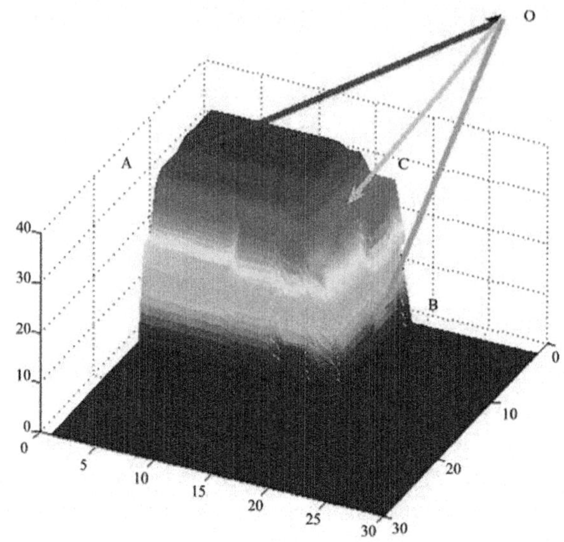

Figure 3: The paths to monotonic indicators requirement locus.

The P_e which meets the effectiveness requirement and has the least budget can be found through optimization. The corresponding duality programming is

$$\text{Max}(E)$$

$$\text{s.t}: C(P_0, P_e) \leq C_K, E = \int_{PI} f(p, P_e)\mathrm{d}p$$

This model will be used to get the maximized capability indicator effectiveness value with the budget being less than C_K.

Time-Oriented Optimization Design

Capability development is usually a time-consuming work. The following optimization model is able to find the path from the original capability point to MIRL with the least time.

$$\text{Min}\left(\text{Max}\left(T_i\left(P_0, P_e\right)\right)\right), 1 \leq i \leq M$$

$$\text{s.t}: E \geq K, E = \int_{PI} f(p, P_e)\mathrm{d}p$$

In this model: $T_i(P_0, P_e)$ represents time-cost function of the ith capability indicator varying from P_0 to P_e and M represents the total number of capability indicators. This model can also be transformed into the following programming model,

$$\begin{aligned}&\mathrm{Min}(\lambda)\\&\text{s.t.}: T_i(P_0,P_e)\le\lambda.,1\le i\le M\\&\quad E\ge K, E=\int_{Pl} f(p,P_e)\mathrm{d}p\end{aligned}$$

Similarly, the dual programming can be changed into

$$\begin{aligned}&\mathrm{Max}(E), E=\int_{Pl} f(p,P_e)\mathrm{d}p\\&\text{s.t}: T_i(P_0,P_e)\le T_K,1\le i\le M\end{aligned}$$

A NOTIONAL EXAMPLE: NUMBER OPTIMIZATION OF MISSILES AVAILABLE

In this section, some models and algorithms mentioned above are used to program missile available numbers of six types so as to meet the mission requirement with the least budget cost.

Problem Context

Ten types of opponent targets (i.e., T1, T2, T3, T4, T5, T6, T7, T8, T9, and T10) are assumed to be destroyed to win the air superiority. Target-attacking missile types are assumed to be type A, type B, type C, type D, type E, and type F.

Because missile production cost is usually very large, so the missile number available of each type is an important capability indicator. The capability indicator optimization design will be carried out in a MIS with six dimensions (i.e., six missile available number indicators). The required number of one certain type missile, when only this type of missile is used to destroy one target of a certain type, is presented in the following table. "No" in the table means that type of missile can not be assigned to attack the corresponding type of target.

If there exists solution meeting the following equality and inequalities group, it means that the missile number available of each type can satisfy the mission requirement.

$$\begin{cases} \sum_{i=1}^{6} \frac{x_{ij}}{R_{ij}} = 1, \ 1 \le j \le 10 \\ \sum_{j=1}^{10} x_{ij} \le W_i, \ 1 \le i \le 6 \\ x_{ij} \ge 0 \end{cases}$$

In the above equation and inequalities group, x_{ij}, R_{ij}, and W_i respectively represent the number of the ith type of missile assigned to the jth type of target, the required number of the ith type missile to destroy jth type target, and the number available of ith type missile.

If the number available of each type of missiles all meet the mission, the requirement measure u is set to 1, otherwise, set to 0. And the function $u = f(W_1, \cdots, W_6)$ is obviously monotonic.

4.2. Capability Indicator Analysis and Optimization Design

The MIS-based capability indicator analysis and optimization design is performed according to the following steps.

1) Construct the MIS, i.e., select all the MIS dimensions and standardize every dimension value range into the range [0,100].
2) Create the monotonic function representing the relation between the capability indicators and requirement measure. Actually, the function $u = f(W_1, \cdots, W_6)$ is already put forward in section 4.1.
3) Apply the partitioning algorithm to acquire the MIRL, i.e., accomplish capability indicators requirement analysis.
4) Use the capability indicators effectiveness model (introduced in section 2.2) to judge whether the present capability meet the requirement or not. If not, go to step5 to perform optimization design, else stop capability indicator analysis and optimization design.
5) Trade off missile number for each type with the optimization model introduced in section 3.1.

Based on the data of Table 1, the MIRL shown in Figure 4 is acquired by the partitioning algorithm. The numbers of six types of missiles (i.e., type A, B, C, D, E, and F) available is respectively assumed to be 14 - 26 number unit, 14 - 26 number unit, 12 - 29 number unit, 14 - 26 number unit, 14 - 26 number unit, and 12 - 29 number unit.

All the numbers of six types of missiles available are assumed to be subject to normal distribution and according to 6σ principle, the corresponding distribution functions can be approximately stated as follows:

$$f_A(N_A)=\frac{1}{\sqrt{(2\pi)2}}e^{\frac{-(N_A-20)^2}{8}},f_B(N_B)=\frac{1}{\sqrt{(2\pi)2}}e^{\frac{-(N_B-20)^2}{8}},f_C(N_C)=\frac{1}{\sqrt{(2\pi)3}}e^{\frac{-(N_C-20)^2}{18}}$$

$$f_D(N_D)=\frac{1}{\sqrt{(2\pi)2}}e^{\frac{-(N_D-20)^2}{8}},f_E(N_E)=\frac{1}{\sqrt{(2\pi)2}}e^{\frac{-(N_E-20)^2}{8}},f_F(N_F)=\frac{1}{\sqrt{(2\pi)3}}e^{\frac{-(N_F-20)^2}{18}}$$

Furthermore, the budget costs of six types of incremental missiles are assumed to be evaluated with the following functions:

$$C(Na)=\begin{cases}25Na & Na\geq 40\\ 15Na & Na<40\end{cases};;;$$

$$C(Nd)=\begin{cases}25Nd & Nd\geq 30\\ 15Nd & Nd<30\end{cases};;.$$

In these functions: $Na, Nb, Nc, Nd, Ne,$ and Nf respectively represent the incremental missile numbers of the type A, B, C, D, E, and F; $C(Na),C(Nb),C(Nc),C(Nd),C(Ne),$ and $C(Nf)$ respectively represent the budget costs of the incremental missiles of type A, B, C, D, E, and F.

If the effectiveness value E (introduced in section 2.2) is required to be larger than 0.9, the optimized incremental numbers of six types of missiles can be acquired through conducting the following programming model.

Table 1: The required number of a certain type missile to destroy a certain type target

target missile	T1	T2	T3	T4	T5	T6	T7	T8	T9	T10
Type A	20	45	12	18	30	6	No	No	No	No
Type B	18	75	18	14	30	8	No	No	No	No
Type C	15	60	12	No	20	No	No	No	No	No
Type D	No	No	No	No	No	No	75	45	24	60
Type E	No	No	No	No	No	No	45	37	24	40
Type F	No	No	No	No	No	No	No	45	20	40

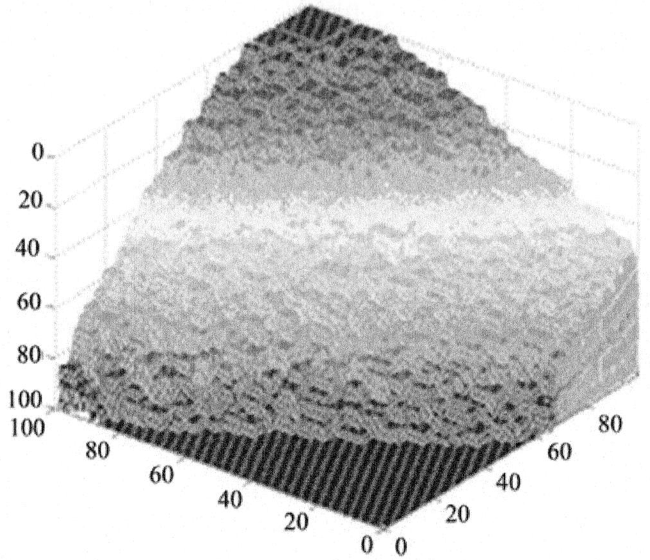

(a)

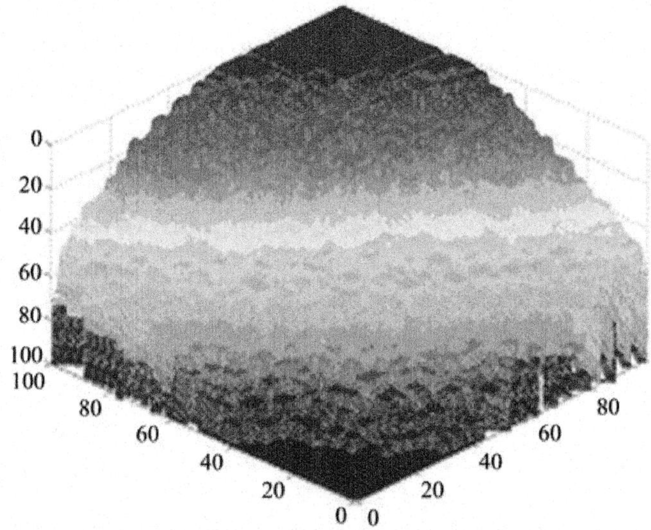

(b)

Figure 4: (a) The MIRL of missile type A, B, and C; (b) The MIRL of missile type D, E, and F.

$$\text{Min}\left(C(Na)+C(Nb)+C(Nc)+C(Nd)+C(Ne)+C(Nf)\right)$$
$$\text{s.t. } E=\int_{PL} f_A(Na+20,2)f_B(Nb+20,2)f_C(Nc+20,3)f_D(Nd+20,2)f_E(Ne+20,2)f_F(Nf+20,3)\mathrm{d}N_A\cdots\mathrm{d}N_F \geq 0.9.$$

In this model: $f_A(Na+20,2), f_B(Nb+20,2), f_C(Nc+20,3), f_D(Nd+20,2), f_E(Ne+20,2)$, and $f_F(Nf+20,3)$ respectively represent the probability density functions of missile numbers of type A, B, C, D, E, and F; $Na+20, Nb+20, Nc+20, Nd+20, Ne+20$, and $Nf+20$ respectively represent the expectation values of six distribution functions; Na, Nb, Nc, Nd, Ne, and Nf respectively represent the incremental missile numbers of type A, B, C, D, E, and F. The initial numbers of six types of missiles are all 20 and variances of six distribution functions are 2 or 3.

Through solving the programming model mentioned above, we get such conclusions that the optimized incremental missile numbers of type A, B, C, D, E, and F should be 28 number unit, 28.2 number unit, 27.9 number unit, 25.7 number unit, 36.5 number unit, and 35.9 number unit and the optimized least budget cost will be 3275 (cost unit).

CONCLUSION AND PERSPECTIVE

This series of papers propose and introduce a novel model, i.e., MVS (monotonic vector space) model, in which there exist some monotonic mappings. Many MVS operations (such as partitioning operation, synthesizing operation, sampling operation, etc.) can be directly used to solve related practical problems, which have been elaborated in the first one of this series papers. Based on MVS, some system capability indicator analysis and optimization models (e.g., requirement analysis model and effectiveness analysis model) are put forward and used to perform capability indicator analysis and optimization design in the second one of this series of papers. The future work for MVS model includes: 1) improving the present operation algorithms and developing other types of operations (such as partitioning operation in non-deterministic discrete or non-deterministic continuous MVS), which is of great significance to practical problems; 2) extending the application of the MVS model in other fields such as imaging processing and multi-object programming, etc.

REFERENCES

1. WSEIAC Report, “AFSC TR-65-2, Vol. 1-3, Final Report of Task Group 2, Preciction Measurement,” 1965.

2. C. L. Hwang, et al., "Multiple Attribute Decision Making," Springger-Verlag, Berlin, 1981.http://dx.doi.org/10.1007/978-3-642-48318-9
3. W. D. Cook and M. A. Kress, "A Multiple-Criteria Composite Index Model for Quantitative and Quantitative Data," European Journal of Operation Research, Vol. 78, No. 3, 1994, pp. 367-379. http://dx.doi.org/10.1016/0377-2217(94)90046-9
4. T. L. Saaty, "The Analytic Hierarchy Process," McGraw-Hill, 1980.
5. V. Bouthonnier and A. H. Levis, "System Effectiveness Analysis of C3 Systems," IEEE Transaction on Systems, Man, and Cybernetics, Vol. 14, No. 1, 1984, pp. 48-54.http://dx.doi.org/10.1109/TSMC.1984.6313268
6. P. K. Davis, "Analytic Architecture for Capabilities-Based Planning, Mission-System Analysis, and Transformation," RAND MR-1513-OSD, 2002.
7. M. C. Bohner, "Computer Graphics for System Effectiveness Analysis," M.S. Thesis, Massachusetts Institute of Technology, Dept. of Electrical Engineering and Computer Science, 1986.
8. X. F. Wu, "SEA Methods and Its Application in C3I System(4)-System Effectiveness Analysis," System Engineering Theory and Practice, Vol. 2, No. 2, 1999, pp. 44-49.
9. B. A. Nixon, "Management of Performance Requirements for Information Systems," IEEE Trans on Software Engineering, Vol. 26, No. 12, 2000..
10. P. H. Cothier and A. H. Levis, "Timeliness and Measures of Effectiveness in Command and Control," In: S. J. Andriole and S. M. Halpin, Eds., Information Technology for Command and Control, IEEE Press, New York, 1991.
11. N. M. Paul, "Evolution Equations of C3I: Toward a New Understanding of Military of Combat," In: Application of Artificial Intelligence to Command & Control Systems IEE Computing Series 13, Peter Peregrinus Ltd..
12. L.Washington, "Effectiveness Analysis of Flexible Manufacturing Systems LIDS-TH-1430," 1985.
13. B. Liu, "Uncertain Programming with Applications," Tsinghua Springer Press, Beijing, 2003.

Chapter 6

USING DEA AND AHP FOR RATIO ANALYSIS

Mohammad Sadegh Pakkar

Faculty of Management, Laurentian University, Sudbury, Canada

ABSTRACT

This research proposes an integrated approach to the Data Envelopment Analysis (DEA) and Ana- lytic Hierarchy Process (AHP) methodologies for ratio analysis. According to this, we compute two sets of weights of ratios in the DEA framework. All ratios are treated as outputs without explicit inputs. The first set of weights represents the most attainable efficiency level for each Decision Making Unit (DMU) in comparison to the other DMUs. The second set of weights represents the relative priority of output-ratios using AHP. We assess the performance of each DMU in terms of the relative closeness to the priority weights of output-ratios. For this purpose, we develop a parametric goal programming model to measure the deviations between the two sets of weights. In- creasing the value of a parameter in a defined range of efficiency loss, we explore how much the deviations can be improved to achieve the desired goals of the decision maker. This may result in various ranking positions for each DMU in comparison to the other DMUs. An illustrated example of eight listed companies in the steel industry of China is used to highlight the usefulness of the proposed approach.

INTRODUCTION

Ratio analysis is a commonly used analytical tool for measuring the relative performance of a Decision Making Units (DMUs) by focusing on one input/ output at a time [1] . In practical applications, the weighted average of a set of individual ratios is applied in order to produce a single measure of performance from various ratios. How- ever, the appropriate assignment of weights in this approach has been an issue of controversy for researchers. Fortunately, the development of modern Operations Research/Management Science (OR/ MS) has provided us with two powerful methods called Data Envelopment

Analysis (DEA) and Analytic Hierarchy Process (AHP) which can be used to derive ratio weights.

DEA is a data-oriented approach for assessing the relative efficiency of Decision Making Units (DMUs) that use multiple inputs to produce multiple outputs. The standard DEA models are formulated using absolute nu- merical input and output data [2] . In order to use ratio data in the DEA framework, ratio-based DEA models were developed as a combination of DEA methodology and ratio analysis [3] - [5] .

However, the weights provided by (ratio-based) DEA models are often unrealistic. Assigning the extremely large or small weights to certain (ratio) data may be undesirable for some Decision Makers (DMs) because it may categorize a DMU as efficient for unlikely weight combinations. Hence, there is a point to argue that the relative priority of (ratio) data should be taken into account in efficiency assessments.

On the other hand, AHP is a multi-attribute decision-making method that can reflect a priori information about the relative priority of (ratio) data in efficiency assessments. AHP usually involves three stages for gene- rating weights. These stages are the decomposition into a hierarchy, comparative judgments, and synthesis of priorities [6]. AHP, however, seldom is used as a stand-alone tool in recent years. Rather, it is combined with other OR/MS models [7]. The most common approach is the imposition of weight restrictions in DEA models. Referring to the literature, AHP can estimate the bounds of the following restrictions in DEA:

- Absolute weight restrictions. These restrictions directly impose upper and (or) lower bounds on the weights of inputs (outputs) using AHP [8].
- Relative weight restrictions. These restrictions limit the relationship between the weights of inputs (outputs) using AHP [9] - [13] .
- Virtual weight restrictions. A single virtual input (output) is defined as the weighted sum of all inputs (out- puts). We refer to the proportion of each component of such sum as the "virtual weight" of an input (output). These restrictions limit virtual weights using AHP [14] [15] .
- Restrictions on input (output) units. These restrictions impose bounds on changes of inputs (outputs) while the relative importance of such changes is computed using AHP [16] .

There are a number of other methods that do not necessarily apply additional restrictions to a DEA model. Such as converting the qualitative data in DEA to the quantitative data using AHP [17] - [24] , ranking the effi- cient/inefficient units in DEA models using AHP in a two stage process [25] - [27] , weighting the efficiency scores obtained from DEA using AHP [28] , weighting the inputs and outputs in the DEA structure [29] - [31] , constructing a convex combination of weights using AHP and DEA [32] and estimating the missing data in DEA using AHP [33] .

Recently the author has applied AHP weights as weight bounds into a two-level DEA model [34] and a common set of weights (CSW) model [35] . Using a similar approach, this research applies AHP in the ratio-based DEA framework to develop a new integration between DEA, AHP and ratio analysis theory.

METHODOLOGY

This research has been organized to proceed along the following stages (Figure 1):

1) Computing the efficiency of each DMU using ratio-based DEA Model (3) without explicit inputs. The computed efficiencies are applied in Model (7).
2) Computing the priority weights of output-ratios for all DMUs using AHP, which impose weight bounds into Model (7).

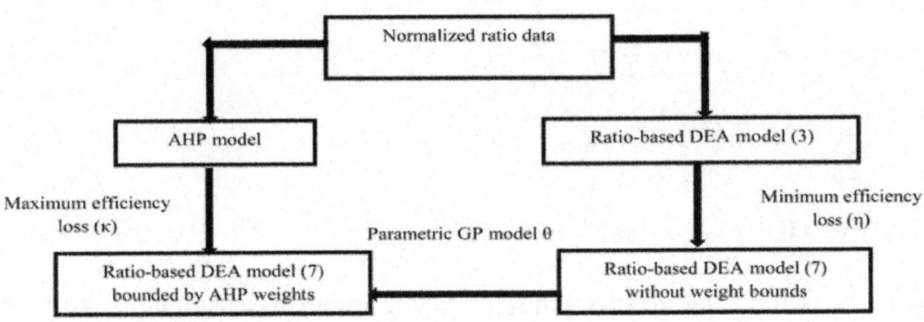

Figure 1: An integrated approach to DEA, AHP and ratio-analysis.

3) Obtaining an optimal set of weights for each DMU using ratio-based DEA Model (7) (minimum efficiency loss η).
4) Obtaining an optimal set of weights for each DMU using Model (7) bounded by AHP (maximum efficiency loss κ). Note that if the AHP weights are added to (7), we obtain Model (9).
5) Measuring the performance of each DMU in terms of the relative closeness to the priority weights of out- put-ratios. For this purpose, we develop a parameter goal programming model. Increasing a parameter in a de- fined range of efficiency loss, we explore how much a DM can achieve its goals. This may result in various ranking positions for a DMU in comparison to the other DMUs.

A Ratio-Based DEA Model

A ratio-based DEA model can be formulated similar to a classical DEA model without explicit inputs [36] [37]. Since the data in a DEA model are in different units of measurements and some output-ratios are negative, the normalized counterparts of output-ratios are computed as follows [38]:

$$y_{rj} = \frac{\hat{y}_{rj} - \hat{y}_{r(\min)}}{\hat{y}_{r(\max)} - \hat{y}_{r(\min)}}, \quad \hat{y}_{r(\min)} = \min\{\hat{y}_{r1}, \hat{y}_{r2}, \cdots, \hat{y}_{rn}\}, \quad \hat{y}_{r(\max)} = \max\{\hat{y}_{r1}, \hat{y}_{r2}, \cdots, \hat{y}_{rn}\}, \tag{1}$$

where $\hat{y}_{rj}$ is the output-ratio $r\,(r = 1,2,\cdots,s)$ for DMU $j\,(j = 1,2,\cdots,n)$. $\hat{y}_{r(\min)}$ and $\hat{y}_{r(\max)}$ are the mini- mum and maximum output-ratios r for all DMUs, respectively. Assume now DMUs have the unit input $i\,(i = 1,2,\cdots,m)$. Then the fractional CCR-DEA model with normalized (ratio) data can be developed as follows [39] :

$$\begin{aligned}
&\text{Max } E_k = \frac{\sum_{r=1}^{s} u_r y_{rk}}{\sum_{i=1}^{m} v_i} \\
&\text{s.t. } \frac{\sum_{r=1}^{s} u_r y_{rj}}{\sum_{i=1}^{m} v_i} \le 1, \quad j = 1,2,\cdots,n, \\
&\qquad u_r, v_i > 0, \quad r = 1,\cdots,s, i = 1,\cdots,m,
\end{aligned} \tag{2}$$

where E_k is the relative efficiency of DMU under assessment. k is the index for the DMU under assessment where k ranges over $1,2,\cdots,n$. v_i and u_r are the weights of input $i\,(i = 1,2,\cdots,m)$ and output-ratio $r\,(r = 1,2,\cdots,s)$. The first set of constraints assures that if the computed weights are applied to a group of n DMUs, $(j = 1,2,\cdots,n)$, they do not attain an efficiency of larger than 1. The second sets of constraints indicate the non-negative conditions for the model variables. Model (2) can be converted to the following linear programming model:

$$\begin{aligned}
&\text{Max } E_k = \sum_{r=1}^{s} u_r y_{rk} \\
&\text{s.t. } \sum_{r=1}^{s} u_r y_{rj} \le 1, \quad j = 1,2,\cdots,n \\
&\qquad u_r > 0, \quad r = 1,\cdots,s.
\end{aligned} \tag{3}$$

Model (3) looks like a DEA model without inputs that combine the standard DEA methodology with ratio analysis.

The AHP Formulation

There are basically three steps for considering decision problems by AHP namely: 1) Decomposition; 2) Pair- wise Comparisons and Judgment Matrix; and 3) Synthesis [6] .

Step 1: Decomposition. This step includes decomposition of the decision problem into elements according to their common characteristics and the formation of a hierarchical model having different levels. In this study, the AHP hierarchical model has three levels: problem objective, criteria and sub-criteria. The problem objective is to prioritize the output-ratios, the criteria are the categories of output-ratios, and the sub-criteria are various out- put-ratios that are organized into these categories (Figure 2).

Step 2: Pairwise Comparisons and Judgment Matrix. In this step, a decision maker makes a pairwise comparison matrix of different criteria, denoted by A with the entries of a_{ho} $(h = o = 1,2,\cdots,S)$. The comparative importance of criteria is provided by the decision maker using a rating scale. Saaty [6] recommends using a 1 - 9 scale. In a similar way, a pairwise comparison matrix can be made to compare the importance of each sub-criterion with respect to a criteria. This matrix is denoted by B with the entries of b_{hlt} $(l = t = 1,2,\cdots,s')$.

Step 3: Synthesis. In this step, the AHP method obtains the priority weights of criteria by computing the eigenvector of matrix A (Equation (4)), $w = (w_1, w_2, \cdots, w_S)^T$, which is related to the largest eigenvalue, $\lambda_{\max}$.

$$Aw = \lambda_{\max} w \quad (4)$$

In a similar way, the priority weights of sub-criteria under each criterion is obtained by computing the Eigen- vector of matrix B (Equation (5)), $e_h = (e_{h1}, e_{h2}, \cdots, e_{hs'})^T$.

$$Be_h = \lambda_{\max} e_h \quad (5)$$

To determine whether or not the inconsistency in a comparison matrix is reasonable the random consistency ratio, $C.R.$, can be computed by the following equation:

$$C.R. = \frac{\lambda_{\max} - N}{(N-1)R.I.} \quad (6)$$

where $R.I.$ is the average random consistency index and N is the size of a comparison matrix.

A Ratio-Based DEA Model Using AHP

We develop our formulation based on a simplified version of the generalized distance model (see for example [40] and [41]) for output-ratios. Let $E_k^* \ (k = 1, 2, \cdots, n)$ be the best attainable efficiency level for the DMU un- der assessment, calculated from Model (3). We want the efficiency $E_k(u)$, calculated from the vector of weights $u = (u_1, \cdots, u_s)$ to be closest to E_k^*. Our definition of closest is that the largest distance is at its minimum. Hence we choose the form of the minimax model: $\min_u \max_k \{E_k^* - E_k(u)\}$ to minimize a single deviation which is equivalent to the following linear model:

$$\begin{aligned}
&\text{Min } \eta \\
&\text{s.t. } E_k^* - \sum_{r=1}^{s} u_r y_{rk} \leq \eta, \\
&\sum_{r=1}^{s} u_r y_{rj} \leq E_j^*, \ j = 1, \cdots, n \\
&\eta \geq 0, \ u_r > 0, \ r = 1, \cdots, s.
\end{aligned} \tag{7}$$

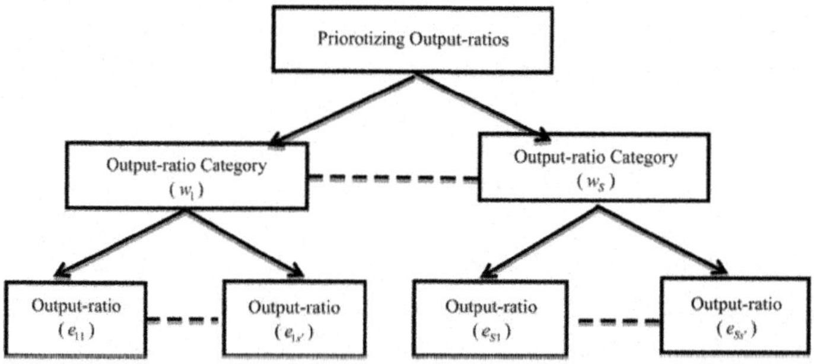

Figure 2: The AHP hierarchical model for prioritizing ratios.

Model (7) identifies the minimum efficiency loss η (eta) needed to arrive at an optimal set of weights. The first constraint ensures that each DMU loses no more than η of its best attainable efficiency, E_k^*. The second set of constraints satisfies that the efficiencies of all DMUs are less than or equal to their upper bound of E_j^*. It should be noted that for each DMU, the minimum efficiency loss η = 0. Therefore, the optimal set of weights obtained from Model (7) is exactly similar to that obtained from Model (3).

To obtain the weight bounds for the weights of output-ratios in Model (7), this study aggregates the priority weights of two different levels in AHP as follows:

$$\bar{u}_r = w_h e_{hl}, \ \sum_{h=1}^{S} w_h = 1, \ \sum_{l=1}^{s'} e_{hl} = 1, \text{ and } r = (h-1)(s') + l, \tag{8}$$

where w_h is the priority weight of criterion $h\,(h = 1,\cdots,S)$ and e_{hl} is the priority weight of sub-criterion $l\,(l = 1,\cdots,s')$ under criterion h in AHP. Note that $[(h-1)(s')+l]$ converts double index (h,l) in AHP to single index $r\,(r = 1,2,\cdots,s = S \times s')$ in DEA. Therefore, to keep the link between the weights of output-ratios and the corresponding priority weights, we use single index r.

In order to estimate the maximum efficiency loss κ (kappa) necessary to achieve the priority weights of output-ratios for each DMU the following linear program is proposed:

$$\begin{aligned}
&\text{Min } \kappa \\
&\text{s.t.} \quad u_r = \alpha \bar{u}_r, \ r = 1,\cdots,s \\
&\qquad E_k^* - \sum_{r=1}^{s} u_r y_{rk} \le \kappa, \\
&\qquad \sum_{r=1}^{s} u_r y_{rj} \le E_j^*, \quad j = 1,\cdots,n \\
&\qquad \kappa \le 1, \ \alpha, \kappa \ge 0, \ u_r > 0, \ r = 1,\cdots,n.
\end{aligned} \tag{9}$$

The first set of constraints changes the priority weights of output-ratios to weights for the new system by means of a scaling factor α. The scaling factor α is added to avoid the possibility of contradicting constraints leading to infeasibility or underestimating the relative efficiencies of DMUs [42] .

A Parametric Goal Programming Model

In this stage, we develop a parametric goal programming model that can be solved repeatedly to generate the various sets of weights for the discrete values of the parameter θ, such that $0 \le \theta \le \kappa$. The purpose of the model is to minimize the total deviations from the priority weights of output-ratios with a city block distance measure. Choosing such a distance measure, each deviation is being equally weighted subject to the following constraints:

$$\begin{aligned}
\text{Min} \quad & Z_k(\theta) = \sum_{r=1}^{s} \left(d_r^+ + d_r^-\right) \\
\text{s.t.} \quad & u_r - d_r^+ + d_r^- = \alpha \bar{u}_r, \\
& E_k^* - \sum_{r=1}^{s} u_r y_{rk} \le \theta, \\
& \sum_{r=1}^{s} u_r y_{rj} \le E_j^*, \ j = 1, \cdots, n \\
& d_r^+, d_r^- \ge 0, \ u_r > 0, r = 1, \cdots, s
\end{aligned} \tag{10}$$

where d_r^+ and d_r^- are the positive and negative deviations from the priority weight of output-ratio $r \, (r = 1, \cdots, s)$ for DMU_k. The first set of equations indicates the goal equations whose right-hand sides are the priority weights of output-ratios adjusted by a scaling variable.

Because the range of deviations computed by the objective function is different for each DMU, it is necessary to normalize it by using relative deviations rather than absolute ones [43] . Hence, the normalized deviations can be computed by:

$$\Delta_k(\theta) = \frac{Z_k^*(0) - Z_k^*(\theta)}{Z_k^*(0)} \tag{11}$$

where $Z_k^*(\theta)$ is the optimal value of the objective function for $0 \le \theta \le \kappa$. We define $\Delta_k(\theta)$ as a measure of closeness which represents the relative closeness of each DMU to the weights obtained from Model (9) in the range [0, 1]. Increasing the parameter (θ), we improve the deviations between the two systems of weights obtained from Models (7) and (9) which may lead to different ranking positions for each DMU in comparison to the other DMUs. It should be noted that in a special case where the parameter $\theta = \kappa = 0$, we assume $\Delta_k(\theta) = 1$.

A NUMERICAL EXAMPLE: THE FINANCIAL PERFORMANCE ASSESSMENT

In this section we present the application of the proposed approach to assess the financial performance of eight listed companies in the steel industry of China. The companies' financial ratios, adopted from [44] , is presented in Table 1.

Table 2 depicts the hierarchical structure of financial ratios and the corresponding priority weights in the AHP model as constructed in [44] . There are five categories of financial ratios at the criteria level. Each one includes three different ratios at the sub-criteria level.

Solving Model (7) for the DMU under assessment with normalized ratio data, we obtain an optimal set of weights with minimum efficiency loss, $\eta = 0$. This implies that the measure of relative closeness to the AHP weights for the DMU under assessment is $\Delta_k(\eta) = 0$.

On the other hand, solving Model (9) for the DMU under assessment, we adjust the priority weights of out- put-ratios obtained from AHP in such a way that they become compatible with the weights' structure in the ratio- based DEA models.

Table 3 presents the optimal weights of output-ratios as well as its scaling factor for the most efficient per- former, DMU5.

It should be noted that the priority weights of AHP (Table 2) used for incorporating weight bounds on the weights of output-ratios are obtained as $\bar{u}_r = u_r / \alpha = w_h e_{hl}$ while $r = (h-1)(3) + l$.

Table 1: Financial market data (ratios) for eight listed companies in China's steel industry

Outputs (Ratios)	DMU1	DMU2	DMU3	DMU4	DMU5	DMU6	DMU7	DMU8
$\hat{y}_1$	0.544	0.622	0.673	0.737	0.659	0.610	0.616	0.768
$\hat{y}_2$	0.520	0.182	0.494	0.403	0.345	0.641	0.407	0.430
$\hat{y}_3$	1.152	0.459	0.911	0.644	1.147	1.098	0.905	0.710
$\hat{y}_4$	5.416	7.042	4.070	8.867	11.813	6.075	5.714	9.489
$\hat{y}_5$	1.423	1.061	1.138	1.664	2.281	2.568	1.470	1.784
$\hat{y}_6$	−0.177	−0.307	0.187	0.924	−0.551	−0.218	−0.186	0.016
$\hat{y}_7$	0.016	0.071	0.021	0.018	0.007	0.004	0.003	0.005
$\hat{y}_8$	0.023	0.075	0.024	0.030	0.015	0.011	0.005	0.009
$\hat{y}_9$	0.043	0.371	0.053	0.075	0.042	0.018	0.012	0.020
$\hat{y}_{10}$	0.044	0.129	0.306	0.155	0.004	0.026	0.165	0.176
$\hat{y}_{11}$	−1.834	−1.877	−3.242	−2.724	−3.759	−5.427	−5.816	−5.009
$\hat{y}_{12}$	−0.372	−0.270	0.076	−0.250	−0.474	−0.374	−0.436	−0.490
$\hat{y}_{13}$	13.229	9.357	6.869	24.910	43.117	16.830	15.505	26.141
$\hat{y}_{14}$	4.337	3.529	2.235	4.193	6.437	2.590	4.312	3.682
$\hat{y}_{15}$	15.258	7.221	18.143	8.060	20.385	39.809	60.400	35.571

Table 2: The AHP hierarchical model*

Objective level	Criteria level	Sub-criteria level	
Prioritizing the financial ratios	Solvency $w_1 = 0.1142$	Current asset ratio $(\hat{y}_1)$	$e_{11} = 0.6144$
		Asset-liability ratio $(\hat{y}_2)$	$e_{12} = 0.2684$
		Current ratio $(\hat{y}_3)$	$e_{13} = 0.1172$
	Asset management and operating ability $w_2 = 0.0553$	Inventory turnover $(\hat{y}_4)$	$e_{21} = 0.3089$
		Total assets turnover $(\hat{y}_5)$	$e_{22} = 0.5816$
		Cash per share $(\hat{y}_6)$	$e_{23} = 0.1095$
	Profitability $w_3 = 0.2302$	Profit margin on sales ($(\hat{y}_7)$	$e_{31} = 0.5396$
		Net profit margin on total assets $(\hat{y}_8)$	$e_{32} = 0.2969$
		Net profit margin on current assets $(\hat{y}_9)$	$e_{33} = 0.1634$
	Development capability $w_4 = 0.1581$	Total assets growth rate $(\hat{y}_{10})$	$e_{41} = 0.1571$
		Profit growth rate $(\hat{y}_{11})$	$e_{42} = 0.5936$
		Sales growth rate $(\hat{y}_{12})$	$e_{43} = 0.2493$
	Profitability to the shareholders $w_5 = 0.4422$	Earnings per share $(\hat{y}_{13})$	$e_{51} = 0.3325$
		Net assets per share $(\hat{y}_{14})$	$e_{52} = 0.1396$
		Price to earnings ratio $(\hat{y}_{15})$	$e_{53} = 0.5278$

* w_h = Criterion $h(h=1,2,3,4,5)$, e_{hl} = Sub-criterion $l(l=1,2,3)$ under criterion h, $\hat{y}_r$ = Output-ratio $r(r=1,2,\cdots,5\times3)$ while $r=(h-1)(3)+l$.

Table 3: Optimal weights of output-ratios of Model (9) for DMU5

u_1	u_2	u_3	u_4	u_5	u_6	u_7	u_8
0.160	0.070	0.031	0.039	0.073	0.014	0.283	0.156
u_9	u_{10}	u_{11}	u_{12}	u_{13}	u_{14}	u_{15}	α
0.086	0.057	0.214	0.090	0.335	0.141	0.532	2.2788

The maximum efficiency loss for the DMU under assessment to achieve the corresponding weights in Model (9) is equal to κ (Table 4). As a result, the measure of relative closeness to the AHP weights for the DMU un- der assessment is $\Delta_k(\kappa)=1$.

Going one step further to the solution process of the parametric goal programming model in (10) we proceed to the estimation of total deviations from the AHP weights for each DMU while the parameter θ is $0 \le \theta \le \kappa$ in (10). Table 5 represents the ranking position of each DMU based on the minimum deviation from the priority weights of output-ratios for $\theta = 0$. It should be noted that in a special case where the parameter $\theta = \kappa = 0$, we assume $\Delta_k(\theta)=1$.

Table 5 shows that DMU5 is the best performer in terms of efficiency and the relative closeness to the priority weights of output-ratios in comparison to the other DMUs. Nevertheless, increasing the value of θ from 0 to κ has two main effects on the performance of the other DMUs: improving the degree of deviations and reducing the efficiency. This, of course, is a phenomenon one expects to observe frequently. The graph of $\Delta(\theta)$ versus θ, as shown in Figure 3, is used to describe the relation between the relative closeness to the priority weights of output-ratios and efficiency loss for each DMU. This may result in different ranking positions for each DMU in comparison to the other DMUs (see Appendix A).

Table 4: Minimum and maximum efficiency loss

Efficiency Loss	DMU1	DMU2	DMU3	DMU4	DMU5	DMU6	DMU7	DMU8
η	0	0	0	0	0	0	0	0
κ	0.34	0.066	0.311	0.146	0	0.276	0.143	0.119

Table 5: The ranking position of each DMU based on the minimum distance to priority weights of output-ratios

θ	DMU1	DMU2	DMU3	DMU4	DMU5	DMU6	DMU7	DMU8
$Z^*(0)$	0.6744	0.0693	0.3767	0.1462	0	0.3863	0.1884	0.2122
Rank	8	2	6	3	1	7	4	5

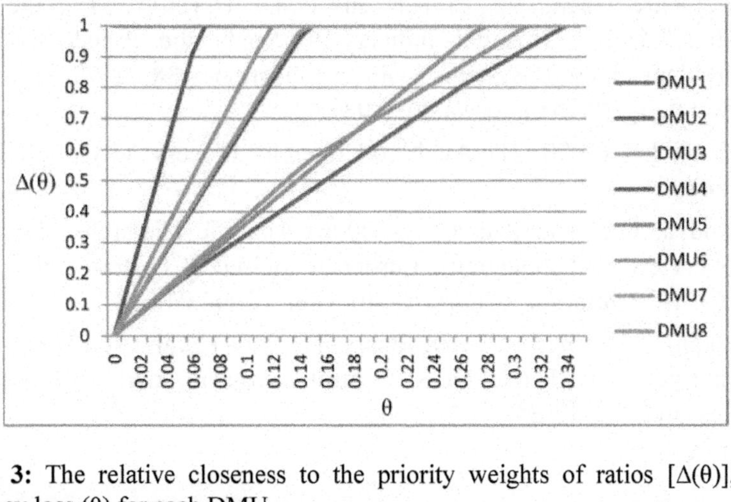

Figure 3: The relative closeness to the priority weights of ratios [Δ(θ)], versus efficiency loss (θ) for each DMU.

CONCLUSION

We develop an integrated approach based on DEA and AHP methodologies for ratio analysis. We define two sets of weights of ratios in the DEA framework. All ratios are treated as outputs without implicit inputs. The first set represents the weights of output-ratios with minimum efficiency loss. The second set represents the corresponding priority weights using AHP with maximum efficiency loss. We assess the performance of each DMU in comparison to the other DMUs based on the relative closeness of the first set of weights to the second set of weights. Improving the measure of relative closeness in a defined range of efficiency loss, we explore the various ranking positions for the DMU under assessment in comparison to the other DMUs.

REFERENCES

1. Sigaroudi, S. (2010) Incorporating Ratios in DEA—Applications to Real Data. Master Thesis, University of Toronto, Toronto.
2. Cooper, W.W., Seiford, L.M. and Zhu, J. (2004) Handbook on Data Envelopment Analysis. Kluwer Academic Publishers, Norwel.
3. Wu, D., Liang, L., Huang, Z. and Li, S. (2005) Aggregated Ratio Analysis in DEA. International Journal of Information Technology & Decision Making, 4, 369-384. http://dx.doi.org/10.1142/S0219622005001593
4. Despic, O., Despic, M. and Paradi, J.C. (2007) DEA-R: Ratio-Based Comparative Efficiency Model, Its Mathematical Relation to DEA and Its Use in Applications. Journal of Productivity Analysis, 28, 33-44. http://dx.doi.org/10.1007/s11123-007-0050-x
5. Wei, C.K., Chen, L.C., Li, R.K. and Tsai, C.H. (2011) Using the DEA-R Model in the Hospital Industry to Study the Pseudo-Inefficiency Problem. Expert Systems with Applications, 38, 2172-2176. http://dx.doi.org/10.1016/j.eswa.2010.08.003
6. Saaty, T.S. (1980) The Analytic Hierarchy Process. McGraw-Hill, New York.
7. Vaidya, O.S. and Kumar, S. (2006) Analytic Hierarchy Process: An Overview of Applications. European Journal of Operational Research, 169, 1-29. http://dx.doi.org/10.1016/j.ejor.2004.04.028
8. Entani, T., Ichihashi, H. and Tanaka, H. (2004) Evaluation Method Based on Interval AHP and DEA. Central European Journal of Operations Research, 12, 25-34.

9. Lee, A.H.I., Lin, C.Y., Kang, H.Y. and Lee, W.H. (2012) An Integrated Performance Evaluation Model for the Photovoltaics Industry. Energies, 5, 1271-1291. http://dx.doi.org/10.3390/en5041271
10. Liu, C.M., Hsu, H.S., Wang, S.T. and Lee, H.K. (2005) A Performance Evaluation Model Based on AHP and DEA, Journal of the Chinese Institute of Industrial Engineers, 22, 243-251. http://dx.doi.org/10.1080/10170660509509294
11. Takamura, Y. and Tone, K. (2003) A Comparative Site Evaluation Study for Relocating Japanese Government Agencies out of Tokyo. Socio-Economic Planning Sciences, 37, 85-102. http://dx.doi.org/10.1016/S0038-0121(02)00049-6
12. Tseng, W., Yang, C. and Wang, D. (2009) Using the DEA and AHP Methods on the Optimal Selection of IT Strategic Alliance Partner. Proceedings of the 2009 International Conference on Business and Information (BAI 2009), 6, 1-15.
13. Kong, W. and Fu, T. (2012) Assessing the Performance of Business Colleges in Taiwan Using Data Envelopment Analysis and Student Based Value-Added Performance Indicators. Omega, 40, 541-549. http://dx.doi.org/10.1016/j.omega.2011.10.004
14. Premachandra, I.M. (2001) Controlling factor weights in data envelopment analysis by Incorporating decision maker's value judgement: An approach based on AHP, Journal of Information and Management Science, 12, 1-12.
15. Shang, J. and Sueyoshi, T. (1995) Theory and Methodology—A Unified Framework for the Selection of a Flexible Manufacturing System. European Journal of Operational Research, 85, 297-315. http://dx.doi.org/10.1016/0377-2217(94)00041-A
16. Lozano, S. and Villa, G. (2009) Multiobjective Target Setting in Data Envelopment Analysis Using AHP. Computers & Operations Research, 36, 549-564. http://dx.doi.org/10.1016/j.cor.2007.10.015
17. Azadeh, A., Ghaderi, S.F. and Izadbakhsh, H. (2008) Integration of DEA and AHP with Computer Simulation for Railway System Improvement and Optimization. Applied Mathematics & Computation, 195, 775-785. http://dx.doi.org/10.1016/j.amc.2007.05.023
18. Ertay, T., Ruan, D. and Tuzkaya, U.R. (2006) Integrating Data Envelopment Analysis and Analytic Hierarchy for the Facility Layout Design in Manufacturing Systems. Information Sciences, 176, 237-262. http://dx.doi.org/10.1016/j.ins.2004.12.001

19. Jyoti, Banwet, D.K. and Deshmukh, S.G. (2008) Evaluating Performance of National R&D Organizations Using Integrated DEA-AHP Technique. International Journal of Productivity and Performance Management, 57, 370-388. http://dx.doi.org/10.1108/17410400810881836
20. Korpela, J., Lehmusvaara, A. and Nisonen, J. (2007) Warehouse Operator Selection by Combining AHP and DEA Methodologies. International Journal of Production Economics, 108, 135-142. http://dx.doi.org/10.1016/j.ijpe.2006.12.046
21. Lin, M., Lee, Y. and Ho, T. (2011) Applying Integrated DEA/AHP to Evaluate the Economic Performance of Local Governments in China. European Journal of Operational Research, 209, 129-140. http://dx.doi.org/10.1016/j.ejor.2010.08.006
22. Ramanathan, R. (2007) Supplier Selection Problem: Integrating DEA with the Approaches of Total Cost of Ownership and AHP. Supply Chain Management, 12, 258-261. http://dx.doi.org/10.1108/13598540710759772
23. Yang, T. and Kuo, C. (2003) A Hierarchical AHP/DEA Methodology for the Facilities Layout Design Problem. European Journal of Operational Research, 147, 128-136. http://dx.doi.org/10.1016/S0377-2217(02)00251-5
24. Raut, R.D. (2011) Environmental Performance: A Hybrid Method for Supplier Selection Using AHP-DEA. International Journal of Business Insights & Transformation, 5, 16-29.
25. Ho, C.B. and Oh, K.B. (2010) Selecting Internet Company Stocks Using a Combined DEA and AHP Approach. International Journal of Systems Science, 41, 325-336. http://dx.doi.org/10.1080/00207720903326902
26. Jablonsky, J. (2007) Measuring the Efficiency of Production Units by AHP Models. Mathematical & Computer Modelling, 46, 1091-1098. http://dx.doi.org/10.1016/j.mcm.2007.03.007
27. Sinuany-Stern, Z., Mehrez, A. and Hadad, Y. (2000) An AHP/DEA Methodology for Ranking Decision Making Units. International Transactions in Operational Research, 7, 109-124. http://dx.doi.org/10.1111/j.1475-3995.2000.tb00189.x
28. Chen, T.Y. (2002) Measuring Firm Performance with DEA and Prior Information in Taiwan's Banks. Applied Economics Letters, 9, 201-204. http://dx.doi.org/10.1080/13504850110057947
29. Cai, Y.Z. and Wu, W.J. (2001) Synthetic Financial Evaluation by a Method of Combining DEA with AHP. International Transactions in Operational Research, 8, 603-609. http://dx.doi.org/10.1111/1475-3995.00336

30. Feng, Y.J., Lu, H. and Bi, K. (2004) An AHP/DEA Method for Measurement of the Efficiency of R&D Management Activities in Universities. International Transactions in Operational Research, 11, 181-191. http://dx.doi.org/10.1111/j.1475-3995.2004.00450.x

31. Kim, T. (2000) Extended Topics in the Integration of Data Envelopment Analysis and the Analytic Hierarchy Process in Decision Making. Ph.D. Thesis, Agricultural & Mechanical College, Louisiana State University, Baton Rouge.

32. Liu, C. and Chen, C. (2004) Incorporating Value Judgments into Data Envelopment Analysis to Improve Decision Quality for Organization. Journal of American Academy of Business, Cambridge, 5, 423-427.

33. Saen, R.F., Memariani, A. and Lotfi, F.H. (2005) Determining Relative Efficiency of Slightly Non-Homogeneous Decision Making Units by Data Envelopment Analysis: A Case Study in IROST. Applied Mathematics and Computation, 165, 313-328. http://dx.doi.org/10.1016/j.amc.2004.04.050

34. Pakkar, M.S. (2014) An Integrated Approach Based on DEA and AHP. Computational Management Science.http://dx.doi.org/10.1007/s10287-014-0207-9

35. Pakkar, M.S. (2012) An Integrated Approach to the DEA and AHP Methodologies in Decision Making. In: Charles, V. and Kumar, M., Eds., Data Envelopment Analysis and Its Applications to Management, Cambridge Scholars Publishing, Newcastle upon Tyne, 136-149.

36. Liu, W.B., Zhang, D.Q., Meng, W., Li, X.X. and Xu, F. (2011) A Study of DEA Models without Explicit Inputs. Omega, 39, 472-480. http://dx.doi.org/10.1016/j.omega.2010.10.005

37. Mozaffari, M., Gerami, J. and Jablonsky, J. (2014) Relationship between DEA Models without Explicit Inputs and DEA-R Models. Central European Journal of Operations Research, 22, 1-12. http://dx.doi.org/10.1007/s10100-012-0273-4

38. Sahoo, B.K. and Meera, E. (2008) A Comparative Application of Alternative DEA Models in Selecting Efficient Large Cap Market Securities in India. International Journal of Management Perspectives, 1, 62-75.

39. Charnes, A., Cooper, W.W. and Rhodes, E. (1978) Measuring the Efficiency of Decision Making Units. European Journal of Operational Research, 2, 429-444. http://dx.doi.org/10.1016/0377-2217(78)90138-8

40. Mavi, R.K., Mavi, N.K. and Mavi, L.K. (2012) Compromise Programming for Common Weight Analysis in Data Envelopment Analysis. American Journal of Scientific Research, Issue 45, 90-109.

41. Hashimoto, A. and Wu, D.A. (2004) A DEA-Compromise Programming Model for Comprehensive Ranking. Journal of the Operation Research Society of Japan, 47, 73-81.

42. Podinovski, V.V. (2004) Suitability and Redundancy of Non-Homogeneous Weight Restrictions for Measuring the Relative Efficiency in DEA. European Journal of Operational Research, Amsterdam, 154, 380-395. http://dx.doi.org/10.1016/S0377-2217(03)00176-0

43. Romero, C. and Rehman, T. (2003) Multiple Criteria Analysis for Agricultural Decisions. 2nd Edition, Elsevier, Amsterdam.

44. Li, H.Y., Zhang, C. and Zhao, D. (2010) Stock Investment Value Analysis Model Based on AHP and Gray Relational Degree. Management Science and Engineering, 4, 1-6.

A: The Measure of Relative Closeness to the Priority Weights of Ratios [Δ(θ)] Verses Efficiency Loss [θ] for Each DMU

θ	DMU1	DMU2	DMU3	DMU4	DMU5	DMU6	DMU7	DMU8
0	0.000000	0.000000	0.000000	0.000000	1.000000	0.000000	0.000000	0.000000
Rank	N/A	N/A	N/A	N/A	1	N/A	N/A	N/A
0.01	0.035504	0.151440	0.038492	0.068376	1.000000	0.036198	0.069881	0.088642
Rank	8	2	6	5	1	7	4	3
0.02	0.071009	0.302879	0.076984	0.136753	1.000000	0.072396	0.139761	0.177284
Rank	8	2	6	5	1	7	4	3
0.03	0.106513	0.454319	0.115476	0.205129	1.000000	0.108594	0.209642	0.264573
Rank	8	2	6	5	1	7	4	3
0.04	0.142017	0.605759	0.153967	0.273505	1.000000	0.144792	0.279522	0.347212
Rank	8	2	6	5	1	7	4	3
0.05	0.176984	0.757199	0.192459	0.341882	1.000000	0.180990	0.349403	0.429852
Rank	8	2	6	5	1	7	4	3
0.06	0.206628	0.908638	0.230951	0.410258	1.000000	0.217188	0.419283	0.512491
Rank	8	2	6	5	1	7	4	3
0.07	0.236273	1.000000	0.269443	0.478634	1.000000	0.253386	0.489164	0.595131
Rank	8	1	6	5	1	7	4	3
0.08	0.265918	1.000000	0.307935	0.547011	1.000000	0.289583	0.559044	0.677770
Rank	8	1	6	5	1	7	4	3

0.09	0.295563	1.000000	0.346427	0.615387	1.000000	0.325781	0.628925	0.760410
Rank	8	1	6	5	1	7	4	3
0.1	0.325208	1.000000	0.384919	0.683763	1.000000	0.361979	0.698805	0.843049
Rank	8	1	6	5	1	7	4	3
0.11	0.354853	1.000000	0.423411	0.752140	1.000000	0.398177	0.768686	0.925689
Rank	8	1	6	5	1	7	4	3
0.12	0.384498	1.000000	0.461902	0.820516	1.000000	0.434375	0.838566	1.000000
Rank	8	1	6	5	1	7	4	1
0.13	0.414143	1.000000	0.500394	0.888892	1.000000	0.470573	0.908447	1.000000
Rank	8	1	6	5	1	7	4	1
0.14	0.443788	1.000000	0.537662	0.957269	1.000000	0.506771	0.978327	1.000000
Rank	8	1	6	5	1	7	4	1
0.15	0.473432	1.000000	0.572844	1.000000	1.000000	0.542969	1.000000	1.000000
Rank	8	1	6	1	1	7	1	1
0.16	0.503077	1.000000	0.600002	1.000000	1.000000	0.579167	1.000000	1.000000
Rank	8	1	6	1	1	7	1	1
0.17	0.532722	1.000000	0.626545	1.000000	1.000000	0.615365	1.000000	1.000000
Rank	8	1	6	1	1	7	1	1
0.18	0.562367	1.000000	0.653088	1.000000	1.000000	0.651563	1.000000	1.000000
Rank	8	1	6	1	1	7	1	1
0.19	0.592012	1.000000	0.679631	1.000000	1.000000	0.687761	1.000000	1.000000
Rank	8	1	7	1	1	6	1	1
0.2	0.621657	1.000000	0.706174	1.000000	1.000000	0.723959	1.000000	1.000000
Rank	8	1	7	1	1	6	1	1
0.21	0.651302	1.000000	0.732717	1.000000	1.000000	0.760157	1.000000	1.000000
Rank	8	1	7	1	1	6	1	1
0.22	0.680947	1.000000	0.759260	1.000000	1.000000	0.796355	1.000000	1.000000
Rank	8	1	7	1	1	6	1	1
0.23	0.710591	1.000000	0.785803	1.000000	1.000000	0.832552	1.000000	1.000000
Rank	8	1	7	1	1	6	1	1
0.24	0.740236	1.000000	0.812345	1.000000	1.000000	0.868750	1.000000	1.000000
Rank	8	1	7	1	1	6	1	1
0.25	0.769881	1.000000	0.838888	1.000000	1.000000	0.904948	1.000000	1.000000
Rank	8	1	7	1	1	6	1	1
0.26	0.799526	1.000000	0.865431	1.000000	1.000000	0.941146	1.000000	1.000000

Rank	8	1	7	1	1	6	1	1
0.27	0.826889	1.000000	0.891974	1.000000	1.000000	0.977344	1.000000	1.000000
Rank	8	1	7	1	1	6	1	1
0.28	0.851599	1.000000	0.918517	1.000000	1.000000	1.000000	1.000000	1.000000
Rank	8	1	7	1	1	1	1	1
0.29	0.876306	1.000000	0.945060	1.000000	1.000000	1.000000	1.000000	1.000000
Rank	8	1	7	1	1	1	1	1
0.3	0.901014	1.000000	0.971603	1.000000	1.000000	1.000000	1.000000	1.000000
Rank	8	1	7	1	1	1	1	1
0.31	0.925721	1.000000	0.998146	1.000000	1.000000	1.000000	1.000000	1.000000
Rank	8	1	7	1	1	1	1	1
0.32	0.950428	1.000000	1.000000	1.000000	1.000000	1.000000	1.000000	1.000000
Rank	8	1	1	1	1	1	1	1
0.33	0.975135	1.000000	1.000000	1.000000	1.000000	1.000000	1.000000	1.000000
Rank	8	1	1	1	1	1	1	1
0.34	0.999842	1.000000	1.000000	1.000000	1.000000	1.000000	1.000000	1.000000
Rank	8	1	1	1	1	1	1	1
0.35	1.000000	1.000000	1.000000	1.000000	1.000000	1.000000	1.000000	1.000000
Rank	1	1	1	1	1	1	1	1

Chapter 7

ASSESSING THE RELATIVE EFFICIENCY OF HEALTH SYSTEMS IN SUB-SAHARAN AFRICA USING DATA ENVELOPMENT ANALYSIS

Samuel Ambapour

Institute National de la Statistique, Brazzaville, Republic of Congo

ABSTRACT

We assess the relative efficiency of health systems of 35 countries in sub-Saharan Africa using Data Envelopment Analysis. This method allows us to evaluate the ability of each country to transform its sanitary "inputs" into health "outputs". Our results show that, on average, the health systems of these countries have an efficiency score between 72% and 84% of their maximum level. We also note that education and density of population are factors that affect the efficiency of the health system in these countries.

INTRODUCTION

Health is now seen as a component of human capital the same way as education and nutritional status [1] -[4]. According to these authors, everyone has an initial health stock that depreciates with age, but can be maintained or even appreciated by combining individual health care and education, and according to the time available. Moreover, according to a study by the World Bank (World Bank, 1993) at least four reasons support the assertion that a healthy individual is more productive and contributes more to economic growth.

- Health limits the loss of production because of the impact of disease on labor.
- It allows exploiting the natural resources that were largely inaccessible because they are located in infested areas.
- It increases the rate of school attendance and allows children to assimilate better the lessons learned.
- Finally, health frees for other purposes, resources that would have served otherwise to provide care to the sick.

The impact of health on the well-being and overall health of a country probably justifies the huge investments of the states in this area. Indeed, in 1990, global spending on health was evaluated at $1700 million [5] with more than 1000 billion from states; representing 60% of the total. In developing countries (Africa, Asia, Latin America), these costs were estimated at $170 billion, 50% funded by the states.

The role and the importance of health systems in the success of health outcomes are now well established. The issues that remain to investigate are, among other things, why some health systems can be considered more effective than others, and what explain the differences in countries' health systems.

The purpose of this paper is to shed some more light on this issue that, to our knowledge, has received little attention in the literature. This relative paucity of literature on the subject is associated, according to some authors (for example, [6] or [7]), to the challenges posed by the comparison of different health systems because, inter alia, of the following reasons:

1) The definition of health proposed by the World Health Organization (WHO) is, according to [8] , useless for all practical purposes, "a perfect state of complete physical, mental and social well-being, not just the absence of disease or illness".
2) There are many measures of health status (see, for example, [6] or [9]), especially if we compare individual health indicators, such as "health utilities index" that, as argued by [7] , are unfortunately based on functional capacity concepts rather than on performance.

This second argument should be tempered because, since 2000, a bold demarche for developing a composite index measuring the performance of health systems was conducted by the World Health Organization [10] . It's an index that determines the overall performance of a health system based both on the level of progress of each country with respect to a number of objectives and on the distribution of the health conditions in the population. Five criteria are generally used for this purpose: 1) The general health; 2) The distribution of this health condition; 3) The responsiveness of the health system; 4) The distribution of responsiveness; 5) The fairness of financial contributions. Unfortunately, as pointed out by [11] , the quality of this synthetic indicator, as well as composite indexes calculated by the United Nations Development Programme (UNDP) namely, the Human Development Index (HDI) or the Human Poverty Index (HPI), is often questioned by statisticians and economists, both in the mathematical formulation as well as in the reliability of the statistics used.

In this paper, we compare and attempt to provide an explanation on the inefficiencies of health systems of 35 countries in sub-Saharan Africa. Our comparative analysis of health systems is based on the concept of efficiency obtained through Data Envelopment Analysis (DEA). This concept is related to the production function that shall be defined as the technical interrelationship which results in the maximum output for a combination of production factors and a given technology. This is somehow the ability of each country to transform its sanitary inputs in health outputs [12] . Beyond this definition, this function is also conceived as a frontier or a standard of comparison for assessing efficiency. In other words, the health system of a country will be considered efficient when the combination of outputs and inputs is located on the frontier.

Several reasons justify the appropriateness of the DEA in this study.

1) The popularity of this technique in the field of health lies in its ability to take into account the specificities of the sector such as the complexity of the technology (multi-product/multi-factors) and the absence of true price both for the outputs and for the inputs [13] .
2) It's suggested for the analysis of complex or non-profit organizations such as public services. As pointed out by [14] , it's close to the work of Leibenstein of X-inefficiency. Indeed, with DEA we can also characterize an output lying inside the Pareto optimal production frontier by stipulating that the hidden inefficiencies come from two sources: 1) the externalities inherent in the economic system or, more generally, to the political and social environment and 2) the non-apparent production factors or not taken into account by the model and thus related to the company's management [15] .

Note that the DEA method was applied in health sector by many other authors, including [16] - [18] . However, in these applications, the analysis is usually at the micro level, that is to say, at the hospital level. The objective is then to evaluate the performance of a hospital in comparison to others [16] . This comparison is sometimes made depending on the status: not-for-profit versus for profit private organizations [17] [18] . It also happens to compare, according to their seniority, the practice of physicians within the same hospital [19] .

Our study differs from previous at least on two points:

1) Our analysis is at a more macro level since we compare different countries' health systems, and not within the same country. We seek to build an international production frontier in the health sector. For each country, we consider all hospitals as a single production unit.
2) In addition, this study is, to our knowledge, one of the first uses of the DEA method to compare health systems of countries in sub-Saharan

Africa.

The rest of this article is organized as follows: Section 2 is devoted to the presentation of the DEA method. In Section 3, we present our results of the evaluation of the technical efficiency of the health systems of 35 countries of sub-Saharan Africa. Thus evaluated, the efficiency depends on the specific environment of each country. To provide explanatory elements of the efficiency scores of the different countries, we establish a relationship between the level of efficiency and certain strategic or environmental variables. Our concluding remarks are provided in Section 4.

METHODOLOGY

We apply DEA to assess the performance of health systems of 35 countries in sub-Saharan Africa. In this section, we first present the DEA and then describe our data and variables.

The DEA Method

DEA is a non-parametric method initially developed by [20] to evaluate the relative efficiency of the decision making units (DMU) of non-profit institutions, or of the public sector which use a group of similar inputs to produce a group of outputs. The DEA method measures the efficiency of a DMU "o" compared with the set of "n" DMUs in a given sample. The aim is to establish a level of relative efficiency θ (0 £ θ £ 1) for each DMU by comparing its input and output quantities with those of other DMUs.

The efficiency in DEA can be characterized in two ways: the input orientation which supposes a minimization of inputs for a given level of outputs and the output orientation which assumes a maximization of the outputs for a given level of inputs. It's also possible to consider constant or variable returns to scale. Our analysis is based on the input minimization model with the assumption of variable returns to scale.

Indeed, minimizing inputs seems appropriate because:

1) One considers that, as in the case of public services, the services provided by the state to citizens are exogenous.
2) Resource utilization by the countries studied is generally carried out in a difficult budgetary situation.
3) Based on our data, input values are more dispersed than those of outputs. Therefore, minimizing inputs should allow better discrimination of efficiency scores of countries' health system.

Besides, the assumption of variable returns to scale can be justified by the fact that it is more general, but also because of our data. Indeed, it's difficult to identify scale inefficiencies in aggregate data as is the case in this study. See [21] for a full discussion of the DEA methods.

The model we have estimated is formally expressed below. All annotations are adopted from [21] and [22] .

$$\min_{\theta,\lambda} \theta \tag{1}$$

Subject to:

$$\sum_{j=1}^{n} \lambda_j X_{ij} \leq \theta X_{io}, \quad i = 1,2,\text{L} ,m \tag{2}$$

$$\sum_{j=1}^{n} \lambda_j Y_{rj} \geq Y_{ro}, \quad r = 1,2,\text{L} ,s \tag{3}$$

$$\sum_{j=1}^{n} \lambda_j = 1, \quad j = 1,2,\text{L} ,n \tag{4}$$

$$\lambda_j \geq 0 \tag{5}$$

where DMU_o represents one of the "n" DMUs under evaluation. x_{io} and y_{ro} are respectively the i[th]input and the r[th] output of the DMU_o. s = the number of outputs produced by the DMU; m = number of inputs. θ* (minθ) is a scalar which represents the score of the technical efficiency allotted to the unit under evaluation and is interpreted as the coefficient of the production level attained by the latter. λ is a weighting allotted to DMUs which helps to determine the envelope formed by efficient DMUs (θ = 1).

Describing the Sample and the Variables

Our data come from the World Bank database [23] . It covers the 1990-1999 periods and involves 35 countries in sub-Saharan Africa.

With aggregated data as ours, we choose as outputs: life expectancy at birth, infant mortality per thousand births and the mortality rate for children under five. These are also some of the outputs generally considered to calculate composite indices measuring the performance of health systems like that of the World Health Organization (WHO, 2000) or of the UNDP (HDI, HPI).

Regarding inputs, like many other authors [16] [17] [19] [22] , we distinguish between labor inputs and capital inputs. The Labor is measured by the number of doctors per 1000 inhabitants. The capital stock is represented by the number of hospital beds per 1000 inhabitants and health expenditures per capita.

To check the sensitivity of our results, we analyze three specifications of the DEA model obtained by different combinations of inputs and outputs (Table 1). The first two specifications differ only on the outputs. In the first model (DEA1), life expectancy at birth and mortality rates of children under five were selected as outputs. In the second model (DEA2), only infant mortality per thousand births was selected as output. Finally, in the third model (DEA3), we select health expenditures as single input.

In the next section, we present the results obtained from these different DEA specifications.

RESULTS

First, we present the efficiency scores of the three DEA specifications. Next, like several other authors (e.g. [24] and [25]) we use, in a second stage, a Tobit regression to analyze the impact of some exogenous factors on the DEA efficiency scores.

The Efficiency Scores

The efficiency scores are shown in Table 2. They were calculated using DEAP software developed by [26] . The complement to 1 of each efficiency score represents the possible proportional reduction of inputs without any reduction of the output levels. In other words, a country that gets a 90% efficiency score can reduce 10% of its health inputs while maintaining its health outputs at the same level. In light of Table 2, we see that the efficiency scores are sensible to the DEA model specifications. Indeed, with DEA1 and DEA2 models, we observe an average efficiency score around 80%. It's only 72% with DEA3. With DEA1, 14 countries out of 35 reach a maximum efficiency of 100%. Seven countries are declared efficient with DEA2 and only four countries are found efficient with DEA3 model specification. These declared efficient health systems constitute the frontier or the comparison reference for the other countries' health system. It is noted, however, that only two countries are found efficient regardless of the type of DEA model specification. Both are countries located in Southern Africa, namely Zambia and Zimbabwe.

The Determinants of the Efficiency Scores of Countries' Health System

The observed efficiency scores reflect not only management errors, but also the environmental factors of each country. In what follows, we will try to establish a relationship between the efficiency scores and a number of structural variables associated with each country.

Table 1: Inputs and outputs of the three different DEA specifications

	Inputs	Outputs
DEA1	1) Number of doctors per 1000 inhabitants 2) Hospital beds per 1000 inhabitants	1) Life expectancy at birth 2) Mortality rate of children under five
DEA2	1) Number of doctors per 1000 inhabitants 2) Hospital beds per 1000 inhabitants	1) Infant mortality per thousand births
DEA3	1) Health expenditures per capita	1) Life expectancy at birth

Table 2: The efficiency scores

Countries	Model 1 (DEA1)		Model 2 (DEA2)		Model 3 (DEA3)	
	Scores	Rank	Scores	Rank	Scores	Rank
Benin	1.000	1	0.739	20	0.660	22
Botswana	0.377	35	0.453	32	0.440	33
Burkina Faso	0.834	20	0.799	17	0.796	12
Burundi	0.699	29	0.838	13	0.796	12
Cameroon	0.776	23	0.584	30	0.584	27
Central African Republic	0.600	33	0.755	19	0.727	17
Chad	0.793	22	0.803	16	0.766	15
Comoros	1.000	1	0.463	33	0.462	32
Democratic Republic of Congo	0.639	31	0.648	27	0.645	25
Congo	0.609	32	0.675	25	0.675	20
Côte D'Ivoire	0.715	27	0.869	12	0.842	9
Djibouti	0.725	26	0.827	14	0.827	10
Ethiopia	0.659	30	0.874	11	0.789	14
Gabon	0.818	19	0.637	28	0.637	26
Gambia	0.730	25	0.614	29	0.569	29
Guinea	0.709	28	0.780	18	0.728	18
Guinea Bissau	0.850	18	0.963	10	0.963	6
Kenya	0.499	34	0.579	31	0.576	28
Madagascar	1.000	1	0.712	22	0.682	19
Malawi	0.902	16	1 .000	1	1.000	1
Mali	1.000	1	1.000	1	0.911	7

Mauritania	1.000	1	0.710	23	0.669	21
Maurice	1.000	1	0.154	35	0.145	35
Mozambique	0.805	21	1.000	1	1.000	1
Niger	1.000	1	1.000	1	0.898	8
Nigeria	0.739	24	0.671	26	0.652	24
Rwanda	0.867	17	0.983	9	0.981	5
Sao Tome	1.000	1	0.384	34	0.380	34
Senegal	1.000	1	0.725	21	0.550	31
Sudan	1.000	1	0.682	24	0.557	30
Tanzania	0.931	15	0.995	8	0.800	11
Togo	1.000	1	0.811	15	0.658	23
Uganda	1.000	1	1.000	1	0.762	16
Zambia	1.000	1	1.000	1	1.000	1
Zimbabwe	1.000	1	1.000	1	1.000	1
Average score	0.837	-	0.764	-	0.718	-

The literature distinguishes five main categories of factors that could affect inefficiency in the health system of a country ([6] [8] [9] [27] - [29]). These are:

1) Economic variables. These include, among others:

a) The level of economic development as measured by real income per capita calculated assuming purchasing power parity. Indeed, a high income should lead to improved efficiency of the health system. However, it should be noted that the influence of income on health is not as straightforward. It passes through the consumption of goods affecting health (nutrition, hygiene, medical care, education, etc.) The empirical relationship may therefore seem mixed if one also introduces in the regression the variables that characterize the level of consumption of these goods.

b) The extent of poverty and income inequality. Since the poor have limited access to health services, it is expected a positive relationship between inefficiency and the extent of poverty. Similarly, it would be legitimate to think that an unequal income distribution would correspond to a worse health conditions. However, difficult to quantify, the concepts of poverty and inequality are suffering from a lack of universally accepted rigorous definition. Furthermore, there is an abundance of potential indicators for these two related phenomena. As to poverty, because of the lack of satisfactory indicators, either the human poverty index (HPI) or the percentage of the labor force employed in agriculture is used, assuming that the majority of poor are in rural areas. As to inequality in income distributions, the Gini index is often chosen as the relevant explanatory variable.

2) The social and health environment variables. It is assumed that there is a link between the risk of infectious diseases and the quality of the health environment. The frequently used indicators are either the percentage of the population with access to safe water supply or those with access to sanitation services. It is expected a negative correlation between these variables and inefficiency.

3) The parental education. The positive effect of this factor, especially women's education, was emphasized by Caldwell [27] . Indeed, a higher parent education leads to better child nutrition, finer use of health services and greater attention to hygiene. As variables to characterize the instruction, one retains either the literacy rate or the enrollment rate. The most likely hypothesis is that a low level of literacy or schooling is associated with a low efficiency.

4) The demographic variables. In this case, one often uses the density of the population. The expected relationship between this variable and inefficiency is not a priori obvious. For developing countries, particularly in Africa, two other indicators are used: the percentage of the population below 15 years or below five years. The latter is more relevant because the majority of deaths in Africa occur before the age of five years. So there should be a positive relationship between this percentage and inefficiency.

5) The nature of the political regime. According to the UNDP [30], democratic regimes achieve higher health outcomes than dictatorial regimes. One often used variable is the Gastil index of civil liberties and political rights provided by the Freedom House.

Taking into account the availability of data, we estimate the following Tobit model:

$$\text{Ln}\left(1/\text{EFF}_i\right) = \alpha_0 + \alpha_1 \text{HPI}_i + \alpha_2 \text{WATER}_i + \alpha_3 \text{EDU}_i + \alpha_4 \text{DENS}_i + \varepsilon_i \quad (6)$$

where, for country i, EFF = DEA efficiency scores. HPI = the UNPD Human Poverty Index. WATER = percentage of the population without access to safe water supply. EDU = the UNPD Education Index. DENS = density of the population.

The results are shown in Table 3.

We observe from Table 3 that the best results, in terms of the significance of the coefficients, are obtained from DEA2 and DEA3 models. Our comments below relate solely to these two models.

We obtain a surprising result with respect to the economic variable used, which is the poverty index (HPI): an inverse relationship between poverty and inefficiency. This somewhat contradictory result is also obtained if we replace HPI by real GDP per capita. Indeed, we found a positive relationship between GDP and inefficiency: it may be possible to spend abundant resources on health while getting very bad results [30] . A more plausible explanation is that the very poor countries are condemned to manage better their health system because they have no other choice. Besides, starting from zero, the relatively limited resources devoted to health can only seem to improve outputs, such as infant mortality.

The health-related variable has the expected sign, but is not significant. The higher is the percentage of the population without access to improved water sources, the greater is the inefficiency.

Table 3: A Tobit model of the determinants of the efficiency scores of countries' health system

	Model 1 (DEA1)	Model 2 (DEA2)	Model 3 (DEA3)
HPI	−0.076331	−0.19393***	−0.20802***
	(−1.5169)	(−3.8182)	(−4.1145)
WATER	−0.034047	0.018964	0.018635
	(−0.23926)	(1.3866)	(1.3857)
EDU	−6.1215**	−7.8596***	−9.2490***
	(−2.0593)	(−2.7848)	(−3.3037)
DENS	−0.0099076	0.0033546*	0.0035947*
	(−0.51426)	(1.8016)	(1.9381)
CONSTANT	6.7812***	12.023***	13.753***
	(2.0656)	(3.7200)	(4.1224)

***, **, * represent significant coefficients at the 1%, 5% and 10% level respectively.

Our results confirm the role of education as a determinant of efficiency. Indeed, when the level of education rises, the inefficiency decreases.

Finally, with respect to the demographic variable DENS, we found a positive relationship with the inefficiency. A high density leads to an increased inefficiency.

CONCLUSIONS

Using published data covering the 1990-1999 period, this paper assessed the efficiency of 35 sub-Saharan countries' health system using the non-parametric technique of DEA. We found that the average efficiency estimates of the countries health system varied from 72% to 84% depending on the combination of inputs and outputs that were considered.

We go beyond this purely descriptive aspect by seeking to identify the factors that can explain the efficiency scores. Our results show that low density of population and the education level contribute to the efficiency of the health system.

To our knowledge, this is one of the first studies using DEA approach to analyzing the efficiency of the health system of countries in sub-Saharan Africa. Additional studies are necessary to understand better and improve the health system of these countries. For example, it would be interesting to extend this study over a longer period. This extension would analyze several sub-periods in order to see the evolution and performance of the health systems. One might also want to make a comparison with other regions. Successful policies of certain countries or regions can inspire others.

REFERENCES

1. Grossman, G. (1972) On the Concept of Health Capital and Demand of Health. Journal of Political Economy, 80, 224-225. http://dx.doi.org/10.1086/259880
2. Bloom, D. and Canning, D. (2000) The Health and Wealth of Nations. Science, 287, 1207-1209. http://dx.doi.org/10.1126/science.287.5456.1207
3. Adams, P., Hurd, M.D., McFadden, D.L., Merrill, A. and Ribeiro, T. (2003) Healthy, Wealthy, and Wise? Tests for Direct Causal Paths between Health and Socioeconomic Status. Journal of Econometrics, 112, 3-56. http://dx.doi.org/10.1016/S0304-4076(02)00145-8
4. Alsan, M., Bloom, D.E., Canning, D. and Jamison, D. (2007) The Consequences of Population Health for Economic Performance. In: Mills, S., Gibson, L. and Mills, A., Eds., Health, Economic Development and Household Poverty, Routledge, Oxford, 21-39.http://dx.doi.org/10.4324/9780203023570.ch2
5. World Bank (1993) World Development Report: Investing in Health. The World Bank, Washington DC.

6. Brunet-Jailly, J. (1998) Innover dans les Systèmes de Santé: Expériences d'Afrique de l'Ouest. Edition Karthala, Paris, 435 p.
7. Le Gales, C., Buron, C., Costet, N. and Rosman, S. (2001) Développement d'un index d'états de santé pondéré par les utilités en population française: Le Health Utilities Index. Economie & Prévision, 150-151, 71-87. http://dx.doi.org/10.3406/ecop.2001.6350
8. Brunet-Jailly, J. (1990) La Pharmacie Populaire du Mali dans le Contexte de l'Initiative de Bamako. Institut national de recherche en santé publique, Bamako.
9. Duret, E. (1999) Décentralisation, Dépenses Publiques et Mortalité Infantile. Revue d'Economie du Développement, 4, 39-68.
10. WHO (2000) The World Health Report: Health System: Improving Performance. Genève.
11. Brilleau, A. (2004) Les indicateurs Liés à la Mise en Œuvre des Cadres Stratégiques de Lutte contre la Pauvreté. Stateco, 98, 51-72.
12. Bosman, N. and Fecher, F. (1992) Une Étude Comparative de l'Efficacité Technique du Secteur de la Santé au Sein des Pays de l'OCDE. Université de Liège, Liège, CIRIEC, Working Paper 92/08.
13. Leleu, H. and Derveux, B. (1997) Comparaison des Différentes Mesures d'Efficacité Technique: Une Application aux Centres Hospitaliers Français. Économie & Prévision, 129, 101-119. http://dx.doi.org/10.3406/ecop.1997.5866
14. Plane, P. (1997) Efficience Technique et Développement. Introduction. Revue d'Economie du Développement, 3, 3-7.
15. El Asraoui, H., Boussemart, J.P. and Lesourd, J.B. (1999) Rentabilité des Cultures Énergétiques et Frontières d'Efficacité. (Badillo, P.Y. and Paradi, J.C., Eds.)
16. Banker, R.D., Conrad, R. and Strauss, R. (1986) A Comparative Application of Data Envelopment Analysis and Translog Methods: An Illustrative Study of Hospital Production. Management Science, 32, 30-44. http://dx.doi.org/10.1287/mnsc.32.1.30
17. Grosskopf, S. and Valdmanis, V. (1987) Measuring Hospital Performance: A Non-Parametric Approach. Journal of Health Economics, 6, 89-107. http://dx.doi.org/10.1016/0167-6296(87)90001-4
18. Färe, R., Grosskopf, S. and Valdmanis, V. (1989) Capacity, Competition and Efficiency in Hospitals: A Nonparametric Approach. Journal of Productivity Analysis, 1, 123-138.http://dx.doi.org/10.1007/BF00157792

19. Chilingerian, J.A. (1994) Exploring Why Some Physicians' Hospital Practices Are More Efficient: Taking DEA inside the Hospital. In: Charnes, A., Cooper, W.W., Lewin, A. and Seiford, L., Eds., Data Envelopment Analysis: Theory, Methodology, and Applications, Kluwer Press, Norwell, 167-193. http://dx.doi.org/10.1007/978-94-011-0637-5_9

20. Charnes, A., Cooper, W.W. and Rhodes, E. (1978) Measuring the Efficiency of Decision Making Units. European Journal of Operational Research, 2, 429-444.http://dx.doi.org/10.1016/0377-2217(78)90138-8

21. Zhu, J. (2002) Quantitative Models for Performance Evaluation and Benchmarking: Data Envelopment Analysis with Spreadsheets and DEA Excel Solver. International Series in Operations Research and Management Science, Vol. 51, Springer, Berlin.

22. Sedzro, K. and Keita, M. (2009) Assessing the Efficiency of Microfinance Institutions Using Data Envelopment Analysis. Journal of International Finance & Economics, 9, 54-67.

23. World Bank (2002) African Development Indicators. The World Bank, Washington.

24. Hoff, A. (2007) Second Stage DEA: Comparison of Approaches for Modelling the DEA Score. European Journal of Operational Research, 181, 425-435.http://dx.doi.org/10.1016/j.ejor.2006.05.019

25. Sueyoshi, T., Goto, M. and Omi, Y. (2010) Corporate Governance and Firm Performance: Evidence from Japanese Manufacturing Industries after the Lost Decade. European Journal of Operational Research, 203, 724-736.http://dx.doi.org/10.1016/j.ejor.2009.09.021

26. Coelli, T. (1996) A Guide to DEAP Version 2.1: A Data Envelopment Analysis (Computer) Program. CEPA Working Paper 96/08.

27. Flegg, A.T. (1982) Inequality of Income, Illiteracy and Medical Care as Determinants of Infant Mortality in Underdeveloped Countries. Population Studies, 36, 441-458.http://dx.doi.org/10.1080/00324728.1982.10405597

28. Flegg, A.T. (1983) On the Determinants of Infant Mortality in Underdeveloped Countries. International Journal of Social Economics, 10, 38-51.http://dx.doi.org/10.1108/eb013943

29. Brun, J.F. and Mathonat, J. (1997) Les Effets du Financement Extérieur sur le Niveau des Dépenses Publiques d'Education et de Santé dans les Pays en Voie de Développement. Une Analyse Econométrique sur Données de Panel. CERDI, Etudes et documents E.97.15.

30. United Nations Development Program (1993) Human Development Reports. Oxford University Press, New York.

Chapter 8

EVALUATE THE INVESTMENT EFFICIENCY BY USING DATA ENVELOPMENT ANALYSIS: THE CASE OF CHINA

Hualun Zhang, Wei Song, Xiaobao Peng, Xiaoyan Song

School of Public Affairs, University of Science and Technology of China, Hefei, China

ABSTRACT

Although investment is regarded as a key force of China's economic growth, little study has been done to measure China's investment efficiency. The present paper applies the data envelopment analysis (DEA) to Chinese provincial panel data from the year 2003 to 2008 for measuring the investment efficiencies and identifying their trends of Chinese 30 provinces and autonomous regions. A cross-efficient DEA model with considering benevolent formulation is used for providing accurate efficiency scores and completely ranking. The empirical results suggest that the differences of investment efficiency in different regions are distinct but tending to diminish year by year, and the investment efficiencies in some provinces are significantly correlated to their investment rates to the national total investment.

INTRODUCTION

Chinese economy development has gained the great success in the past two decades that attracts worldwide attention. Many researchers confirmed importance of the aggregate investment in the Chinese economic growth and performance see [1-3], which has taken up about 30% of gross domestic product (GDP) see [4,5]. The China's growth pattern is also similarly considered as the investment-growth model in [6]. More and more researchers start to study the investment and investment efficiency, because the importance of investment in Chinese economic growth is widely acknowledged.

In [1,2] admit that aggregate investment plays a important role in China's phenomenal economic growth after estimating China's aggregate economy by consumption function, investment function and production function, respectively. [7] suggests that fixed-capital investment is the most important determinant of China's economic growth, and China follows an investment-driven expansion path in the 1980-1990s. [8] apply the exogeneity framework to investigate empirically the relationship between investment and growth in China. They find that there is a robust and significant relation of capital formation on output growth, suggesting that the fixed investment is a key determinant of China's economic growth. Since investment has been regarded as an important force of China's economic growth, In [4] propose a fixed capital formation to explore and explain the determinants of China's aggregate investment based on a panel data set of 28 provinces and autonomous regions. The empirical results they obtained show the existence of a homogenous equilibrium correction mechanism in China's aggregate investment process and imply the suggestion that it is important to introduce favorable investment incentives in the central and west regions for balancing the economic growth. In [5] divide the aggregate investment in China into business sector investment and government direct investment and separately assess each of them by composing a suitable investment model. Their results suggest that the business sector investment is largely determined by market forces while the government direct investment is found to bear strong planned features. In [9] utilize macroeconometric models to measure the validity of the belief that the Chinese economy still follows largely the investment-led growth paradigm growth. By using the post-1990 annual data, they confirm the relation between investment and economy growth, and suggest that the problem of overinvestment still exists in China.

China's aggregate fixed-asset investment has risen significantly from the year 1987; especially since the year 2002, the growth rate keep higher than 20%; the fixed-asset investment increased much faster than the GDP (see **Table 1**). Such as high investment growth rate might cause the over-investment and investment inefficient, because the output and capital productivity growth lags behind investment see [9,10]. In [11] suggests that decentralisation imposes a variety of interregional barriers to trade and causes the inefficient in resource allocation. By analyzing the variation of total factor productivity and investment returns during the 1990s, In [12] assert that China's allocative efficiency is significantly enhanced by the continuous sectoral shift of labor force. [10] maintain that the phenomenon of over-investment and inefficient investment still exists in China by composing a model that based on the standard capital factor input demand theory with associate measures of allocative inefficiency and technical inefficiency.

Table 1: China's growth rates of aggregate investment and GDP

Year	Aggregate Investment	Growth Rate of AI	GDP	Growth Rate of GDP
1987	3791.8		12050.6	
1988	4753.8	18.5%	15036.8	11.3%
1989	4410.4	11.0%	17000.9	4.1%
1990	4517	4.5%	18718.3	3.8%
1991	5594.5	18.6%	21826.2	9.2%
1992	8080.1	37.6%	26937.3	14.2%
1993	13072.3	22.0%	35260	13.5%
1994	17042.1	15.8%	48108.5	12.6%
1995	20019.3	11.0%	59810.5	10.5%
1996	22913.5	12.7%	70142.5	9.6%
1997	24941.1	10.1%	78060.8	8.8%
1998	28406.2	14.1%	83024.3	7.8%
1999	29854.7	5.2%	88479.2	7.1%
2000	32917.7	9.3%	98000.5	8.0%
2001	37213.5	12.1%	108068.2	7.5%
2002	43499.9	16.1%	119095.7	8.3%
2003	55566.6	27.7%	135174	9.5%
2004	70477.4	26.6%	159586.7	10.1%
2005	88773.6	26.0%	184088.6	10.4%
2006	109998.2	23.9%	213131.7	11.6%
2007	137323.9	24.8%	259258.9	11.9%
2008	172828.4	25.9%	302853.4	9.0%

Although many studies have been done to verify the importance of aggregate investment to China's growth and the existence of China's over-investment, there has little research on measuring the China's investment efficiency detailed from intuitionistic angle. This study seeks to evaluate the efficiency of the aggregate investment of 30 provinces and autonomous regions in China for the year 2003 to 2008 by using a model with relate concepts called cross-efficiency evaluation method. Then, we set out to explore and explain the changing trend of the performances during the six years and provide some manage suggestions for the corresponding decision makers based on the empirical results. The rest of the paper is organised as follows. Section 2 introduces the mathematical programming model used in the study. Application and empirical results are presented in Section 3, while the final section concludes the paper.

METHODOLOGY

Data envelopment analysis (DEA), proposed by [13], has become a popular efficiency evaluation tool and its derivative models have been widely used in many fields to measure the relative efficiency of peer decision making units (DMUs) on the basis of their multiple inputs and outputs. [14] extend DEA to the cross-efficiency evaluation method for identifying the best performing DMUs and ranking them by using cross-efficiency scores that obtained by all of the DMUs. The main idea of cross-efficiency evaluation is to use DEA in a peer evaluation mode, rather than in a self-evaluation mode.

Adopting the conventional nomenclature of DEA, assume that there are n DMUs that are to be evaluated in terms of m inputs and s outputs. We denote the ith input an rth output for DMUj (j = 1, ···, n) as x_{ij}^t (i = 1, ···, m) and y_{rj}^t (r = 1, ···, s) at each time period t, t = 1, ···, T. The efficiency scores of each DMUd (d = 1, ···, n) based on the traditional input-oriented DEA model proposed by [13] can be obtained by the following linear programming model (1) as:

$$\text{Max} \sum_{r=1}^{s} \mu_r y_{rd} = \theta_d^t \left(x_k^t, y_k^t \right)$$

$$s.t. \quad \sum_{i=1}^{m} \omega_i x_{ij}^t - \sum_{r=1}^{s} \mu_r y_{rj}^t \geq 0, \ j = 1, \cdots, n,$$

$$\sum_{i=1}^{m} \omega_i x_{id} = 1,$$

$$\omega_i \geq 0, i = 1, 2, \cdots, m,$$

$$\mu_r \geq 0, r = 1, 2, \cdots, s. \tag{1}$$

$\theta_d^t \left(x_k^t, y_k^t \right)$ is the optimal result of the above model, which shows the performance of DMUd based on a traditional CCR model at the period t. $\omega_{1d}^*, \cdots, \omega_{md}^*, \mu_{1d}^*, \cdots, \mu_{sd}^*$ are the optimal weights obtained by the model (1). Then the cross efficiency of any DMUj (j = 1, ···, n) at period t, using the weight has chosen by DMUd in model (1) can be computed as:

$$E_{dj}^t\left(x_k^t, y_k^t\right) = \frac{\sum_{r=1}^{s} \mu_{rd}^* y_{rj}^t}{\sum_{i=1}^{m} \omega_{id}^* x_{ij}^t}, d, j = 1, \cdots, n. \tag{2}$$

It is clear from **Table 2** that each unit $E_{dj}^t\left(x_k^t, y_k^t\right)$ is the efficiency that DMUd accords to DMUj, if we move along the dth row of the matrix E of cross efficiencies, given the computed weighting scheme described above.

Table 2: A generalized cross-efficiency matrix (CEM)

Rating DMU	Rated DMU				
	1	2	3	...	n
1	$E_{11}^t\left(x_k^t, y_k^t\right)$	$E_{12}^t\left(x_k^t, y_k^t\right)$	$E_{13}^t\left(x_k^t, y_k^t\right)$	...	$E_{1n}^t\left(x_k^t, y_k^t\right)$
2	$E_{21}^t\left(x_k^t, y_k^t\right)$	$E_{22}^t\left(x_k^t, y_k^t\right)$	$E_{23}^t\left(x_k^t, y_k^t\right)$	...	$E_{2n}^t\left(x_k^t, y_k^t\right)$
...	...	...	...	...	...
n	$E_{n1}^t\left(x_k^t, y_k^t\right)$	$E_{n2}^t\left(x_k^t, y_k^t\right)$	$E_{n3}^t\left(x_k^t, y_k^t\right)$	...	$E_{nn}^t\left(x_k^t, y_k^t\right)$
Mean	$\bar{E}_1^t\left(x_k^t, y_k^t\right)$	$\bar{E}_2^t\left(x_k^t, y_k^t\right)$	$\bar{E}_3^t\left(x_k^t, y_k^t\right)$	...	$\bar{E}_n^t\left(x_k^t, y_k^t\right)$

The leading diagonal shows the special case where DMUd rates itself, which means the value of $E_{dd}^t\left(x_k^t, y_k^t\right)$ is exactly equal to the optimal result $\theta_d^t\left(x_k^t, y_k^t\right)$. Then, we average each the columns of cross efficiency matrix (CEM) in **Table 1** to get a mean cross efficiency measure for each DMU. So, the cross-efficiency score of DMUk (k = 1, ···, n) at period t should be obtained by averaging all $E_{dk}^t\left(x_k^t, y_k^t\right)$ (d = 1, ···, n), namely $\bar{E}_k^t\left(x_k^t, y_k^t\right) = \frac{1}{n}\sum_{d=1}^{n} E_{dk}^t\left(x_k^t, y_k^t\right)$ (k = 1, ···, n).

In fact, the weights $\omega_{1d}^*, \cdots, \omega_{md}^*, \mu_{1d}^*, \cdots, \mu_{sd}^*$ obtained by maximizing DMUd' simple efficiency scores may not unique, that possibly reduces the perceived usefulness of cross-efficiency. One remedy suggested is to introduce a secondary objective function to resolve the ambiguity see [14,15]. Following their studies, we conclude a benevolent secondary formulation, which could be utilized to maximize the other DMU's cross-efficiencies based on all the possible weights $\omega_{1d}^*, \cdots, \omega_{md}^*, \mu_{1d}^*, \cdots, \mu_{sd}^*$ obtained by model (1):

$$
\begin{aligned}
\text{Max } & \tilde{E}_{dj}^{t}\left(x_k^t, y_k^t\right) = \sum_{r=1}^{s} \mu_r y_{rj} \\
s.t \quad & \sum_{i=1}^{m} \omega_i x_{il}^t - \sum_{r=1}^{s} \mu_r y_{rl}^t \geq 0, l = 1, \cdots, n, \\
& \sum_{r=1}^{m} \omega_i x_{ij} = 1, \\
& \sum_{i=1}^{m} \omega_i x_{id}^t - E_{dd}^t\left(x_k^t, y_k^t\right) \sum_{r=1}^{s} \mu_r y_{rd}^t = 0, \\
& \omega_i \geq 0, i = 1, 2, \cdots, m, \\
& \mu_r \geq 0, r = 1, 2, \cdots, s.
\end{aligned} \tag{3}
$$

The optimal value $\tilde{E}_{dj}^{t}\left(x_k^t, y_k^t\right)$ of model (2), denoting the maximum efficiency scores of DMUj based on DMUd at period t, could be used to determine the ultimate cross efficiency of each DMUk at period t:

$$
\tilde{E}_k^t\left(x_k^t, y_k^t\right) = \frac{1}{n} \sum_{d=1}^{n} \tilde{E}_{dj}^t\left(x_k^t, y_k^t\right), k = 1, \cdots, n.
$$

APPLICATION

Data Description

There are a great number of input and output variables, whether tangible and intangible, that can describe the influencing factors and results of investment. However, China's economic growth depends very much on the rapid expansion of manufacturing sector see [12]. Taking the factors that have been suggested in the existing literature and data available into consideration, we choose three input variables: Fixed-asset investment at provincial level (X1); Net fixed asset of industry at provincial level (X2); Number of employee of industry at provincial level (X3), while two major economic indicators are selected: Gross Domestic Product (GDP) at the provincial level (Y1); and the Value-added of industry at provincial level (Y2). In this paper, all the input and output variables are selected from Chinese Statistical Yearbook, and six years from 2003 to 2008 are considered.

Efficiency Analysis

The CCR efficiency score of the 30 different provinces in China was firstly assessed using the traditional model (1). From **Table 3**, we can find that there are 7 efficient provinces in 2003, 9 in 2004, 9 in 2005, 11 in 2006, 11 in 2007 and 10 in 2008. Four east region provinces (namely, Beijin, Shanghai, Guangdong and Hainan) and one mid region province (Heilongjiang) keep efficient in all the five years considered, while 17 provinces poorly perform during 2003 to 2008. Besides, we can also conclude from the above table that the performance of east region is better than the mid and west region because of more efficient provinces and higher average efficiency score. However, this CCR efficiency score might not good enough. Firstly, it is calculated by using the weight selected itself, which heavily weights few favorable measures and completely ignores other inputs and outputs in order to maximize its own DEA efficiency. Secondly, we cannot completely rank the performance of the 30 provinces based on these CCR efficiency scores. Finally, for the good-performance provinces (such as Beijing, whose efficiency score keeping 1 during the six years), CCR efficiency scores cannot reflect the performance changes over the six years.

Table 3: CCR efficiency scores of different provinces

Regions	Province	2003	2004	2005	2006	2007	2008
	BJ	1	1	1	1	1	1
	TJ	0.892	0.977	0.985	1	1	1
	HB	0.900	0.914	0.902	0.895	0.912	0.896
	LN	0.833	0.724	0.745	0.743	0.765	0.730
	SH	1	1	1	1	1	1
East regions	JS	0.999	1	0.883	0.907	0.879	0.833
	ZJ	0.927	0.972	0.751	0.744	0.783	0.816
	FJ	1.000	0.961	0.957	0.952	0.923	0.973
	SD	0.997	1	1	1	1	1
	GD	1	1	1	1	1	1
	HAN	1	1	1	1	1	1
	SX	0.750	0.750	0.741	0.725	0.753	0.714
	NMG	0.746	0.749	0.900	0.919	1.000	1.000
	JL	0.848	0.843	0.783	0.799	0.897	0.822
	HLJ	1	1	1	1	1	1
Mid regions	AH	0.909	0.870	0.890	0.885	0.857	0.767
	JX	0.844	0.909	0.963	0.964	0.909	0.708

	HN	0.946	0.937	0.932	0.967	1	1
	HUB	0.771	0.766	0.824	0.834	0.824	0.772
	HUN	1	0.968	1	1	1	1
West regions	GX	1	0.997	1	1	1	0.957
	CQ	0.785	0.846	0.845	0.863	0.825	0.819
	SC	0.788	0.797	0.857	0.848	0.879	0.878
	GZ	0.666	0.679	0.688	0.724	0.737	0.745
	YN	0.951	1	0.967	1	0.949	0.926
	SHX	0.697	0.694	0.700	0.778	0.930	0.807
	GS	0.666	0.700	0.719	0.797	0.783	0.734
	QH	0.580	0.685	0.750	0.771	0.860	0.857
	NX	0.502	0.532	0.586	0.579	0.644	0.651
	XJ	0.921	1	1	1	1	1

Table 4 reports the cross-efficiency scores of the 30 provinces during 2003 to 2008 by utilizing the model (2). As indicated in **Table 2**, no province is complete efficient measured by the cross-efficiency DEA model. To further analyze the cross-efficiency results, we classify all the 30 provinces and autonomous regions into three groups according to the economic region they belong to and compare the average efficiency scores of each group in the six years, see **Figure 1**. Note that the efficiency scores of the east region are higher than the mid, while the efficiency values of mid regions are also higher than the western. Also, more details should be found:

- The changes of efficiency scores of east provinces are very small which express a trend downward.
- The wave of efficiency scores of mid regions is significant.
- The improvement of west provinces are more pronounced.

Table 4: Cross-efficiency scores of provinces during 2003 to 2008

Regions	Province	2003	2004	2005	2006	2007	2008
East regions	BJ	0.827	0.859	0.834	0.808	0.818	0.834
	TJ	0.718	0.768	0.767	0.869	0.868	0.867
	HB	0.778	0.791	0.765	0.772	0.775	0.759
	LN	0.698	0.632	0.639	0.636	0.653	0.632
	SH	0.858	0.841	0.792	0.788	0.816	0.830
	JS	0.783	0.786	0.736	0.748	0.725	0.662
	ZJ	0.692	0.695	0.619	0.605	0.607	0.629
	FJ	0.863	0.802	0.758	0.725	0.672	0.696
	SD	0.757	0.768	0.811	0.790	0.797	0.806
	GD	0.899	0.879	0.816	0.785	0.749	0.727
	HAN	0.850	0.829	0.760	0.712	0.833	0.830

Mid regions	SX	0.600	0.613	0.574	0.540	0.572	0.585
	NMG	0.648	0.617	0.662	0.682	0.733	0.693
	JL	0.750	0.742	0.677	0.664	0.710	0.666
	HLJ	0.888	0.888	0.916	0.912	0.871	0.887
	AH	0.767	0.711	0.728	0.724	0.680	0.617
	JX	0.665	0.649	0.688	0.746	0.713	0.577
	HN	0.785	0.768	0.759	0.797	0.868	0.868
	HUB	0.657	0.635	0.648	0.629	0.680	0.656
	HUN	0.812	0.779	0.816	0.852	0.842	0.799
West regions	GX	0.810	0.763	0.785	0.829	0.799	0.740
	CQ	0.639	0.605	0.608	0.666	0.636	0.615
	SC	0.674	0.673	0.710	0.726	0.752	0.737
	GZ	0.579	0.583	0.578	0.612	0.615	0.628
	YN	0.843	0.831	0.775	0.793	0.763	0.767
	SHX	0.607	0.597	0.617	0.667	0.765	0.685
	GS	0.561	0.580	0.570	0.591	0.631	0.629
	QH	0.424	0.458	0.469	0.482	0.533	0.544
	NX	0.442	0.459	0.495	0.499	0.558	0.552
	XJ	0.635	0.673	0.731	0.773	0.765	0.847

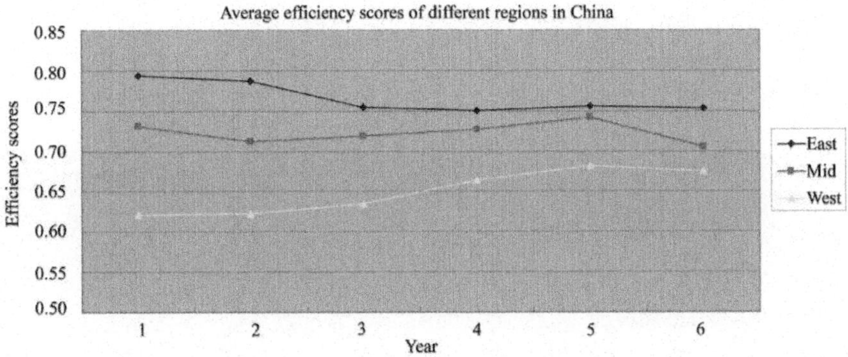

Figure 1: Average efficiency scores of different regions in China.

Besides, the phenomenon of convergence is illustrating in figure 1 too during 2003 to 2008, which might be due to the Chinese central planning to promote the regional productivity equality see [16]. However, that phenomenon might also caused by the over-investing in the east region and infrastructure construction enhancing in the mid and west regions. By comparing the average efficiency scores in different years, we find that the regional efficiency gap evidently decreased from 2003 to 2008, which might indicate that the investment allocation and productivity in different regions across the country are becoming reasonable and equal.

Table 5 reports the complete ranking of different provinces during 2003 to 2008 based on the cross-efficiency scores obtained above. It can be seen that Heilongjiang is best performance province during these six years, because it has held the place of no. 1 for five years since 2004. On the other hand, Qinghai and Ningxia always rank at the end, more work need to be done to improve their performances. Specifically, two provinces, namely, Sichuan and Xinjiang, had increasingly improved their performance order, while two provinces (namely, Anhui and Guangdong) have shown a continuous decreasing trend.

Table 5: Ranking of cross-efficiency scores of provinces during 2003 to 2008

Regions	Province	2003	2004	2005	2006	2007	2008
	BJ	7	3	2	5	6	5
	TJ	16	12	9	2	3	3
	HB	12	8	10	12	10	11
	LN	17	22	22	23	23	21
	SH	4	4	6	9	7	6
East regions	JS	11	9	14	13	17	19
	ZJ	18	17	23	26	27	22
	FJ	3	7	13	16	22	15
	SD	14	11	5	8	9	8
	GD	1	2	4	10	15	14
	HAN	5	6	11	18	5	7
	SX	26	24	27	28	28	27
	NMG	22	23	20	19	16	16
	JL	15	15	19	22	19	18
	HLJ	2	1	1	1	1	1
Mid regions	AH	13	16	16	17	21	25
	JX	20	20	18	14	18	28
	HN	10	13	12	6	2	2
	HUB	21	21	21	24	20	20
	HUN	8	10	3	3	4	9
	GX	9	14	7	4	8	12
	CQ	23	25	25	21	24	26
	SC	19	19	17	15	14	13
	GZ	27	27	26	25	26	24
	YN	6	5	8	7	13	10
West regions	SHX	25	26	24	20	12	17
	GS	28	28	28	27	25	23
	QH	30	30	30	30	30	30
	NX	29	29	29	29	29	29
	XJ	24	18	15	11	11	4

Take XinJiang for example, its ranking order increased from the 24th to 4th during the period 2003-2008. The increase could probably be attributed to the work done by the managers during these six years in improving the efficiency of using the investment. On the other hand, as the ranking order decreased from 1st to 14th during the six years, Guangdong province is an example of the typical provinces which are significant slippage. The decision makers need to do more work to reverse the disadvantageous situation by improving the efficiency of resource use.

Correlational Analysis between the Efficiency Scores and Investment

The above **Figure 2** shows the trend of the cross-efficiency scores of all the 30 provinces during 2003 to 2008. Some provinces express an upward tendency (such as Tianjin, Gansu, Qinghai, Ningxia and Xinjiang), while some others show a downward tendency (such as Jiangsu, Fujian, Guangdong and Anhui). Besides, note that, most of the provinces in west region keep their efficiency ascendant in these years.

Figure 3 gives the curves of ratio of provincial investment to the nationwide total investment during 2003 to 2008. A noticeable feature is the curve trends in figure 3, indicating significant changes of investment. Specifically, the decreasing and increasing tendency could be respectively found in about 10 provinces (such as Beijin, Shanghai, Jiangsu, Zhejiang, Guangdong, Hainan, Guizhou, Qinghai, Ningxia and Xinjiang) and 8 provinces (Hebei, Neimenggu, Jilin, Anhui, Henan, Guangxi, Chongqi and Shanxi). Another noteworthy feature is that no province in middle region shows a decreasing trend on investment ratio in the six years, which illustrates that more investment shift to middle region.

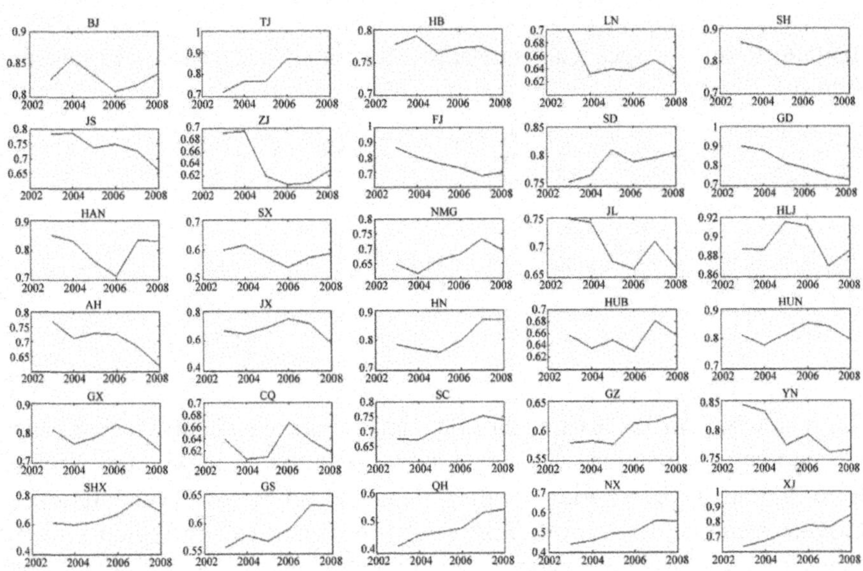

Figure 2: Cross-efficiency scores of provinces during 2003 to 2008.

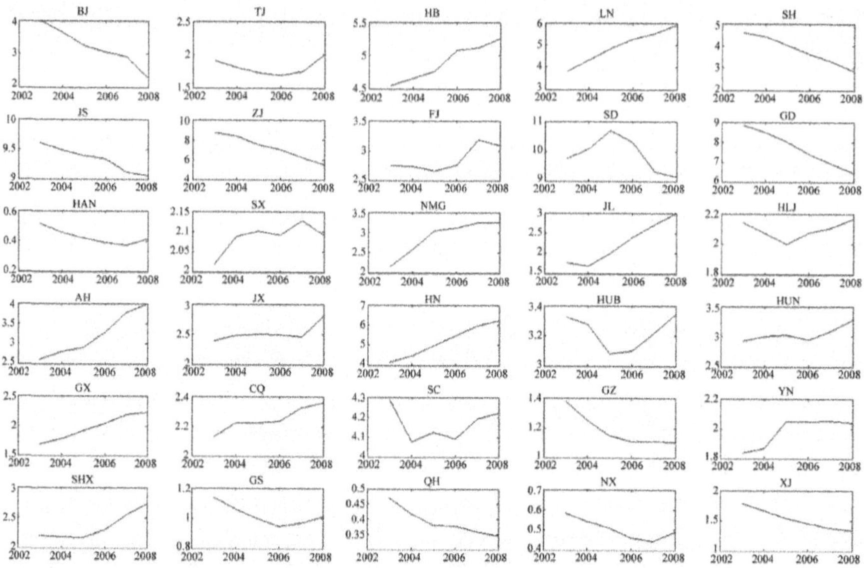

Figure 3: Ratio of provincial investment to the nationwide total investment during 2003 to 2008.

Table 6: Results of Spearman correlation between efficiency score and investment ratio

East region	BJ	TJ	HB	LN	SH	JS	ZJ	FJ	SD	GD	HAN
Spearman Correlation	0.143	−0.429	0.714	−0.486	0.486	0.886	0.543	−0.771	0.086	1.000	0.257
Sig. (2-tailed)	0.787	0.397	0.111	0.329	0.329	0.019	0.266	0.072	0.872	0	0.623
Middle region	SX	NMG	JL	HLJ	AH	JX	HN	HUB	HUN		
Spearman Correlation	−0.771	0.943	−0.657	−0.600	−0.829	−0.257	0.771	0.371	−0.143		
Sig. (2-tailed)	0.072	0.005	0.156	0.208	0.042	0.623	0.072	0.468	0.787		
West region	GX	CQ	SC	GZ	YN	SHX	GS	QH	NX	XJ	
Spearman Correlation	−0.314	0.086	0.314	−0.829	−0.771	0.771	0.600	−1.000	−0.943	−0.943	
Sig. (2-tailed)	0.544	0.872	0.544	0.042	0.072	0.072	0.208	0	0.005	0.005	

It is seen from **Table 6** that the efficiency scores of eight provinces (Jiangsu, Guangdong, Neimenggu, Anhui, Guizhou, Qinghai, Ningxia and Xinjiang) have significant correlation[1] with the investment ratios. Then, these eight provinces can be divided into four categories. The two east region provinces Jiangsu and Guangdong belong to the first category, whose efficiency scores decline with the investment ratios falling. However, on the contrary, the four west region provinces (Guizhou, Qinghai, Ningxia and Xinjiang) improve their performance as the investment ratios decreasing and compose the second category. Lastly, two middle region provinces, Neimenggu and Anhui, are respectively called the third and fourth category because of their different response to the increasing investment ratio. The efficiency score of Neimenggu grow with the increasing of investment ratio, while the performance of Anhui gets worse significantly as investment raise sharply.

CONCLUSIONS

This paper applies a cross-efficiency DEA model to measure the investment efficiency of 30 provinces in China in order to provide a detailed regional analysis of Chinese investment inefficiency. Three inputs (Fixedasset investment, Net fixed asset of industry and Number of employee of industry) and two outputs (GDP and Value-added of industry) and provincial-panel data for the period 2003-2008 are considered in this study. A number of interesting results with important implications are found:

Based on the empirical results, there are very significant differences between the three economic regions in China: the east region is the best performance while the west region is the worst. This is very understandable because of the disparities in the infrastructure. Better investment environment is provided by the east region through years of accumulation, while the mid and west regions do not have a requirement to make full use of the investment.

On the other hand, the assimilation appears in nationwide: the difference of the performances between the three regions is diminishing by comparing the trend of efficiencies during 2003-2008. That phenomenon is not only an expression of the improvement of allocative efficiency but also the inevitable result that economy grows. Moreover, more work need to be done to narrow the gap between different regions in the further.

Investment is driving force for some east province such as Jiangsu and Guangdong that is expressed as the significant positive correlation between the decrease of investment and efficiency, while the over-investment does exist in some west provinces such as Guizhou, Qinghai, Ningxia and Xinjiang because of the significant negative correlation between the investment decreasing and performance improving.

REFERENCES

1. G. C. Chow, “A Model of Chinese National Income Determination,” Journal of Political Economy, Vol. 93, No. 4, 1985, pp. 782-791. doi:10.1086/261330
2. G. C. Chow, “Capital Formation and Economic-Growth in China,” Quarterly Journal of Economics, Vol. 108, No. 3, 1993, pp. 809-867. doi:10.2307/2118409

3. L. X. Sun, "Estimating Investment Functions Based on Cointegration: The Case of China," Journal of Comparative Economics, Vol. 26, No. 1, 1998, pp. 175-191.doi:10.1006/jcec.1997.1501
4. H. Song, Z. Liu and P. Jiang, "Analysing the Determinants of China's Aggregate Investment in the Reform Period," China Economic Review, Vol. 12, No. 2-3, 2001, pp. 227-242. doi:10.1016/S1043-951X(01)00052-9
5. X. He and D. Qin, "Aggregate Investment in People's Republic of China: Some Empirical Evidence," Asian Development Review, Vol. 21, 2004, pp. 99-117.
6. J. Zhang, "Capital Formation, Industrialization and Economic Growth," Economic Research Journal, Vol. 7, 2002, pp. 3-13 (in Chinese).
7. Q. Yu, "Capital Investment, International Trade and Economic Growth in China: Evidence in the 1980-90s," China Economic Review, Vol. 9, No. 1, 1998, pp. 73-84.doi:10.1016/S1043-951X(99)80005-4
8. A. C. C. Kwan, Y. Wu and J. Zhang, "Fixed Investment and Economic Growth in China," Economics of Planning, Vol. 32, No. 1, 1999, pp. 67-79. doi:10.1023/A:1003424418042
9. D. Qin, M. A. Gagas, P. Quising and X. He, "How Much Does Investment Drive Economic Growth in China?" Journal of Policy Modeling, Vol. 28, No. 7, 2006, pp. 751-774.doi:10.1016/j.jpolmod.2006.02.004
10. D. Qin and H. Song, "Sources of Investment Inefficiency: The Case of Fixed-Asset Investment in China," Journal of Development Economics, Vol. 90, No. 1, 2009, pp. 74- 145. doi:10.1016/j.jdeveco.2008.06.001
11. A. Young, "The Razor's Edge Distortions and Incremental Reform in the People's Republic of China," The Quarterly Journal of Economics, Vol. 115, No. 4, 2000, pp. 1091-1136.doi:10.1162/003355300555024
12. J. Zhang, "Investment, Investment Efficiency, and Economic Growth in China," Journal of Asian Economics, Vol. 14, No. 5, 2003, pp. 713-734. doi:10.1016/j.asieco.2003.10.004
13. A. Charnes, W. W. Cooper and E. Rhodes, "Measuring the Efficiency of Decision Making Units," European Journal of Operational Research, Vol. 3, No. 4, 1970, pp. 429- 444.doi:10.1016/0377-2217(79)90229-7
14. T. R. Sexton, R. H. Silkman and A. J. Hogan, "Data Envelopment Analysis: Critique and Extensions," In: R. H. Silkman, Ed., Measuring Efficiency: An Assessment of Data Envelopment Analysis, Jossey-Bass, San Francisco, 1986, pp. 73-105.

15. J. Doyle and R. Green, "Efficiency and Cross Efficiency in DEA: Derivations, Meanings and the Uses," Journal of the Operational Research Society, Vol. 45, 1994, pp. 567- 578.
16. T. Jian, J. Sachs and A. Warner, "Trends in Regional Inequality in China," China Economic Review, Vol. 7, No. 1, 1996, pp. 1-21. doi:10.1016/S1043-951X(96)90017-6

Chapter 9

ANALYSIS OF STUDIES FROM 2000-2010 IN REAL OPTION THEORY AND APPLICATION TO OM

Hui-Chuan Chen

University of Texas at Arlington, US

ABSTRACT

Traditional project investment methods, such as the discounted cash flow (DCF) with a fixed static plan, are no longer sufficient to assist the corporate strategies of seizing opportunities and profitability. The option pricing formula includes a theoretical framework for pricing financial options, assuming that the risk in a financial hedged position is zero, if the option is adjusted continuously in a short position. Hence, the real options revolution arose in response to the dissatisfaction of corporation practitioners with traditional capital budgeting techniques, such as standard discount cash flow. This paper analyzes relevant articles from the "Journal of Operations Management" and "Management Sciences" as related to real options theory in the field of operations management. The goal of this study is to review and identify the gaps in application to real option theory in their studies. Finally, this paper provides suggestions for future researchers.

INTRODUCTION

In today's highly competitive business environment, corporate managers must constantly make decisions in response to the rapidly changing marketplaces, including new product introductions, information systems, research and development, as well as outsourcing of manufacturing. Company strategies must be flexible and agile in order to seize the opportunities and profitability. If managers can learn from their mistakes, and make adjustments to execute different options, companies can more quickly reach a higher profit than their competitors. Yeo and Qiu [1] mentioned that many of these managers now recognize that the traditional project investment methods, especially the discounted cash flow (DCF) based measures are no longer sufficient in the current rapidly changing environment. Traditionally, the DCF method provides a fixed static plan and expects that future plans will not be altered. However, this is no longer a suitable method. This study will provide an overview of real options theory and investigate the relevant literature to determine the gaps which occur when real options are implemented in the operations management field. Furthermore, this study will present some relevant propositions and pertinent questions for managers in decision making processes.

OVERVIEW OF CONVENTIONAL REAL OPTIONS THEORY

Black and Scholes [2] published the option pricing formula in 1973 with a theoretical framework for pricing financial options; specifically, they provide the basic assumption and equation that stock values follow a lognormal distribution progress,

$$\frac{dS}{S} = \mu dt + \sigma dz \tag{1}$$

where S is the price of the stock, μ is the drift rate of S, t is a time in years, σ is the variation of the stock's returns, and z is a Wiener process. Generally speaking, an option provides the holder the right to purchase or sell a share of stock at a specific price. The holders have the right to purchase a stock when it is a "call" option, and they also have the right to sell a stock when it is a "put" option. An organized traded call options started in April 1973, and in June 1977 the trading of put options followed [3]. An American option indicates that the option can be exercised at any time before maturity or the expiration date, whereas a European option can only be exercised on the expiration date. One early pricing theory proposed by Black and Scholes [2] states that at the maturity date and under risk neutral conditions, the price of a European call option has a closed solution. They assume that the risk in the financial hedged position is zero if the option is adjusted continuously in a short position; however, if the position is not constantly modified, the risk is minimal, and is comprised entirely of risk that can be spread by shaping a portfolio of a great number of hedged positions [2]. The mathematical analysis for financial options through diversifying the risks has led to Cox and Ross' [4] risk neutral valuation model. They applied a series of binomial trees or the Monte Carlo method as option valuation techniques to represent future asset values. Cox and Ross further state that investors can expect returns and discount rates in a "realistic" risk free condition. They also pointed out that Black and Scholes' option pricing model relies only on observable variables under a static setting, as opposed to Cox and Ross' option valuation model which focuses on "the aggregate value of the claims against the returns of a firm" [4]. However, the risk neutral valuation may not be realistic for the business environment under many uncertainties and changes.

NEW DIRECTIONS FOR REAL OPTIONS

Trigeorgis [5] indicated that the real options revolution arose in response to the dissatisfaction of corporate practitioners using traditional capital budgeting techniques, such as standard discount cash flow. Yeo and Qiu [1] point out that the real options applications assist managers to gain a broader perspective and opportunities rather than a particular valuation. Thus, various investment scenarios can be seen as groups of options. Furthermore, the main difference between applying real option methods and financial options is that real options are relevant to real resources which are tangible, including machinery, factory, etc. Conversely, a financial asset normally includes stocks and bonds.

Some other distinctions between the financial options theory and the real options involve financial options having a shorter life—such as less than one year of expiration date, whereas real options can be long-lived. Additionally, financial options are fairly simple with a single exercise price, but the exercise price for real options differs from time to time. Yeo and Qiu [1] provide an example that investing in research and development (R&D) produces an option to implement a technology with unknown benefits. If the investment is successful, there is a succeeding option to increase the product line. When the product is at the end of product stage, there is the option to abandon it.

Moreover, the market position of the company may exercise the option to be influenced by a series of options and optimal timing.Amram and Kulatilaka [6] state that applying an options-based approach is not merely the use of a new method of valuation equations and models. In fact, it involves a new approach of structuring strategic decisions. The managers must consider a sequence of strategies such as potential gains for the corporation by moving from position A to position B. It is similar to a decision tress with various options opening down to the decision paths.

Uncovering each step through the real options can be difficult. Real options are different from the financial options because real options are not specifically defined or clearly packaged. However, the options are actually present in many business decisions. Amram and Kulatilaka [6] and Yeo & Qiu [1] indicate some of the hypothetical examples of the most general types of real options: "Timing options, Growth options, Staging options, Exit options, Flexibility options, Operating options, and Learning options."These common types of real options assist the companies to increase the scale by enhancing the upside potentials profits, without raising the downside of the risks. The next section will present some relevant articles which discuss the application of real options theory in the field of operations management.

LITERATURE REVIEW OF REAL OPTIONS APPLICATIONS IN OM

Journal of Operations Management and Management Science are two well-established journals in the field of operations management research. The following literature review presented in this paper will focus on the relevance of real options applications within these two journals as related to practicing real options in OM and the application of simulation methods.

Application of Real Options in the OM

When corporations try to alleviate the risk, the real options can be demonstrated with financial and nonfinancial hedging for risk management strategies. Boyabatlı and Toktay [7] indicate that real options are applied as operational hedging instruments. Much operational hedging has been demonstrated in a diversity of fields such as finance, strategy, operations management, and international business. Operational hedging comprises a major part of firm-level risk management decisions which show that firms actually exercise operational hedges in managing their risks. Specifically, Weiss and Maher [8] discuss how financial and operational hedging impacts the airline industry. After the events of September 11, 2001, airline industries realized the importance of managing financial distress under undesirable conditions and alleviating risk. Specifically, Weiss and Maher's research included nine U.S airlines with data covering 44 quarters from 1990 to 2000 examining the impact of the firms' performance under uncertainty on operational hedging to financial hedging. The attributes for operations hedging are represented by fleet diversification, load factor, lease, and domestic and the attributes for financial hedging are represented by fuel hedging, cash, financial leverage. The results show that the airlines involved in operational hedging can better respond to unfavorable events to reduce risks. Thus, in the case of the airline industry, financial hedging instruments (i.e., fuel price protection) are not as powerful as operations hedging.

Amram and Kulatilaka [6] address the flexibility options when demand is uncertain for new products; however, forecasts imply that sales targets would be reached across two continents. Thus, the managers should decide whether to build a single plant in one continent or two plants on two continents. The flexibility option will be taken into account when the value of the option outweighs the costs saved by only building one plant. In another study,

Jack and Raturi [9] infer that volume flexibility assists the handling of the aggregate demand uncertainty. Volume flexibility allows the company to change production upwards and downwards within broad limits. However, the implementation of flexibility in capabilities may not be easily exercised or accurately measured. Therefore, modular product design permits firms to buffer processes with a list of "modules" in which the span of processes can be buffered resulted in increasing demand uncertainties. Additionally, modular product design and a list of "modules" improve the volume flexibility by generating options for the firm which did not exist before. Their research applied the findings into a survey of 140 valid business respondents to see how each firm employs its resources to accomplish volume flexibility.

The results indicated that both small and large firms depend on overtime source as a key shortterm option of volume flexibility; however, small firms are more efficient at using inventory and capacity buffers (short-term sources) of volume flexibility to react to variations in demand while large firms are better stand on taking competitive advantage through the long-term sources (such as supply chain networks and outsourcing arrangements) of volume flexibility. The findings further suggest that volume flexibility for short and long-term sources has a positive impact on both delivery and financial performance. da Silveria [10] describes the challenge for many operations seeking flexibility without any negative effect on expenditures, quality, delivery, or performance. The flexibility model which included 285 manufacturers of fabricated metal products, machinery, and equipment from 14 countries revealed that flexibility competence could be built by structuring simplicity and discipline in manufacturing. da Silveria further stated that simplicity without lowering the number of options offered to the firm should deliver a rationalized process that is easier to adapt and reconfigure to altering requirements; likewise, discipline, as opposed to "stiffening procedures and skills," will enable a firm to response to changes in the marketplace, while promoting improved practices and work processes.

The results provided the flexibility improvements have a positive relationship between simplicity and discipline in manufacturing. Especially, the relationships were stronger in high volume processes than in low volume processes. Sawhney [11] applied flexibility simultaneously between the reactive and proactive approach to assist managers with their daily operational decision making among the supply chain systems. The reactive approach is addressed when organizations deal with different types of uncertainty which extend from the up and down to the basic task within the firm. As for proactive approaches, it has been argued that flexibility can be used proactively to generate competitive advantages for a firm.

Thus, the area of flexibility is hierarchical, and it bi-directs from up to downstream or vice-versa within a single firm. Moreover, the flexibility under application of reactive and proactive is in a sequential approach which provides additional options for managers planning strategies across the supply chain to create value for the customers and the firm Furthermore, Pagell and Krause [12] have considered the relationship between uncertainty and flexibility and between flexibility and performance. However, their study presented doubts regarding the earlier results. Pagell and Krause replicated the model of Swamidass and Newell [13] and focused on surveys but found no relationship between increased uncertainty and increased flexibility under cross-industry sample of manufacturing firms. Additionally, there is little supporting evidence when higher levels of flexibility in uncertain environments were associated with higher levels of performance. Thus, it was suggested that a more thorough research study should involve industry and business strategy.

Under mass customization, customers are highly involved in specifying the product, whereas manufacturers can produce high volumes of products. In other words, customers can purchase a customized product without sacrificing economies of scale from the cost of a mass production item. Additionally, during mass customization, the manufacturers must attend to each customer's specifications in product design. Duray et al. [14] recommended that the manufacturers employ a modular design to achieve manufacturing efficiencies to approximate the standardization of mass production. Their study found that utilizing the modularity in the later phases of production may improve performance for mass customizers. Moreover, the results showed that when mass customizers approach mass production, profitable scales of economy and better financial performance can be achieved. Dilts and Pence [15] studied whether the role of decision maker—project manager or executive sponsor— would impact the termination a public sector project. Different functional managers have different project perceptions while an individual manager interprets the projects differently; therefore, this will affect how the managers' decisions whether to continue or terminate the projects. Moreover, other reasons to terminate the projects include how much effort and money the managers have devoted, how rapid and uncertainty the external environment changes, or how big or small the scale of the projects are. Also, there are tendencies for individual decision makers to add more resources (sunk costs) already consumed in a failed project. The final finding from Dilts and Pence's study indicated that two key factors have statistically significant difference between executive and project managers.

Executives think that variance in overall project complexity and in time to completion are less significant than the project managers. In other words, project managers (when compared to executives) have a higher tendency to terminate a project which is running overtime. Folta, Delmar, and Wennberg [16] applied the hybrid entrepreneurship to study how individual's decision to opt for self-employment. They defined hybrid entrepreneurs as individuals who take on self-employment while holding a primary job in some other organization.

The authors described that the hybrid entrepreneurship is most likely to experience the entrepreneurial waters before becoming fully immersed into self-employment. Through this experience, individuals can learn the venture's potential advantages whether they can adapt to the self-employment surroundings. When individuals are less confident, this mode of entry might assist them to limit their sunk costs while they collect evidence to better understand the unknown capabilities. Furthermore, a real option may be characterized under small-scale entry with hybrid entrepreneurship to invest heavily when early returns are showing and to retire if they are not.

This situation can also be explained by determining when the switching or opportunity costs are high; the individual might delay being fully self-employed. Research has shown that the hybrids often leave wage work and join self-employment when there is a positive signal about performance prospects; however, the hybrids may leave their self-employment while there is a negative signal. Otherwise, individuals might stay in hybrid status until there is a clear signal. This leads to the conclusion drawn that hybrid entrepreneurship is attractive due to its avoidance of switching costs such as losing retirement benefits, healthcare, or seniority status to maintain the flexibility and option value related to delay entrepreneurial access. The next section will provide some simulation designs that may occur under real options applications.

Simulation Method Applications in OM

Managers are often concerned about making tough decisions that require choosing among various competing designs within the firms' manufacturing, supply chain, or service delivery system. Simulations can be a part of methods to assist the managers for determining alternative designs. However, the time needed to do the simulation or which given simulation results to implement is another important decision for the managers. Thus, if the goal is to choose the high-level system design with maximize expected net present value (NPV), the manager may encounter more simulations to decrease the uncertainty or delay in project implementation by applying more simulations. Ranking and selection procedures which provide a preferred level of statistical evidence in best performance are one of the frequent approaches for selecting a finite set of simulated systems. Additionally, Chick and Gans [17] stated that the ranking and selection methods attempt to minimize the mean number of simulations necessary to achieve a preferred level of statistical evidence for appropriate selection. However, it was emphasized that the statistical significance might not be the same as financial significance; therefore, when simulation results and system performance are used as financial measures, the maximization of expected NPV can be a more appropriate objective. The simulations are based on managers having prior confidence regarding the distribution of the NPV of these. Eventually, the authors designed their study to respond to the issue under simulation with financial measures such as marginal cost under optimization control to apply operational decisions.

Gamba and Fusair [18] stated that real options theory presents an ordinary framework to achieve value creation from modular design. Furthermore, they explain that the modularization process is a detailed description which can be defined by a number of parameters and their associations. Additionally, they point out that a module is determined by a group of strongly interconnected factors that are typically independent from the factors of other modules. Similarly, Baldwin and Clark [19] indicated that module design is to exhibit modularity in design under a complex system.

Table 1: Literature review summary

Journal	Author(s)	Valid Data	Topic	Results
JOM	Weiss & Maher (2009)	11 US Airlines	Operational vs. financial hedging	Operational hedging vehicles are more powerful in protecting firms than using financial instruments.
JOM	Sawhney (2006)	64 Printed Circuit board plants	Flexibility vs. reactive and proactive	Flexibility under application of reactive and proactive is in a sequential approach which provides additional options across the supply chain.
JOM	de Silveira (2006)	285 manufacturers from 14 countries	Flexibility vs. simplicity & discipline	Flexibility improvements have a positive relationship between simplicity and discipline in manufacturing.
JOM	Dilts &Pence (2006)	55 worked for National Institute of Justice public projects	Executives & Project managers' role in termination projects	Project managers are more likely to terminate a project that is running overtime than are executives.
JOM	Pagell & Krause (2004)	252 members of the Institute of Supply Management (ISM)	Flexibility vs. uncertain & performance	The results show that there is no evidence to support higher levels of flexibility in uncertain environments when associated with higher levels of performance.
JOM	Jack& Raturi (2002)	Three case studies & 140 survey	Volume Flexibility & Performance	Volume flexibility for short and long-term options has a positive impact on both delivery and financial performance.
JOM	Duray *et al.* (2000)	194 manuf. Plants	Mass Customization & Modular Design	When mass customizers approach mass production, they can reach economies of scale and better financial performance items.
MS	Folta, Delmar & Wennberg (2010)	45,000 Swedish men	Hybrid entrepreneurship vs. complete immersion in self-employment	Hybrid entry is preferred to self-employment entry with more capable, lower switching costs, and less self-employment experience.
MS	Chick and Gans (2009)	Simulation	Max. NPV vs. Min. replications	It links financial measure to optimal control of simulation experiments that are designed to inform operational decisions.
MS	Gamba & Fusari (2009)	Simulation	Valuing Modularity vs. real option	Based on modularization in the design of a system for capital budgeting decisions.
MS	Kumar & Turnbull (2008)	None	Optimal patenting & licensing of financial innovations	A parsimonious framework is developed to assist the managers whether to patent under consideration for a financial institution.

If the modularity segments can be constructed independently, the design will interconnect to support the whole. Furthermore, Gamba and Fusair considered six operators (splitting, substitution, augmenting, excluding, inversion, and porting) which can be defined as options were chosen to describe the evolution from a nonmodular design to a modular design. Thus, there is a need to link modularity and real option theory to practice. In addition, Gamba and Fusair combined Baldwin and Clark's six modular operators to implement the simulations with each operator. Thus, the purpose for these operators is intended to create value, while the natural valuation application depends on claim analysis employed to optional investment decisionswhich comprisereal options theory.

Financial institutions, which are also applicable for real options, often develop different types of innovations in financial services and products. In order to protect these innovations, financial managers have the option to decide whether to obtain patents, patents and licensing, or none at all. Kumar and Turnbull [20] mentioned that without such measures, the innovating institution has no legal right to seek a judgment if there is imitation. Therefore, while larger financial firms generally patent, nonfinancial institutions file for more patents compared to financial firms. Please see Table 1.Literature Review Summary.

IDENTIFYING THE GAPS IN THE LITERATURE & PROPOSITIONS

Based on the literature review above, this section identifies the lack of real world applications for option theory. Weiss and Maher [8] clearly applied real options (hedging) to understand how airlines manage to alleviate the risk. However, while many firms incorporate flexibility into their strategic planning, the studies do not specifically apply real options theory into the research. For example, both Jack and Raturi [9] and da Silveria [10] used flexibility for their research on volume productions. However, they did not consider whether building flexibility increases the overall corporation value. Thus, these studies could have included Mello, Parsons, and Triantis [21]'s flexibility study in sourcing its production to hedge exchange rate risk in financial markets. The degree of flexibility for production is directly link with the firm's financial policy. Therefore, the following propositions are established:

- *Proposition* 1: Higher degree of production flexibility is positively related to a firm's willingness to hedge its financial assets

For example, Sawhney [11] studied options for both suppliers and customers' flexibility to reduce the manufacturing uncertainty. He provided a transformation framework of flexibility for his model. Also, he discusses how flexibility assists the companies to reduce costs. Nevertheless, the manner in which these different flexibilities affect the financial impact of the firms is not addressed. Pagell and Krause [12] studied the flexibility and firms' performance, and they tried to evaluate the flexibility with some financial impact such as growth in sales and returns for the firms. However, they did not find a significant relationship between the flexibility and performance. Duray et al. [14] utilized mass customization with modular design to achieve high performance. Since a positive relationship between mass customization and financial performance was observed, I contend that:

- *Proposition* 2: The higher the mass customization utilized in the firm, the more likely it is to build stronger flexibility in the firm and its financial performance.

Many research studies have presented different options and provide simulations to investigate which option can offer the highest value for the corporations. These options and simulations are usually predetermined by the managers who make the final decision, resulting in the highest value of simulation. Thus, these options are quite difficult to utilize in real world examples. For example, Boyabatlı and Toktay [7] provided some real options among operational flexibility. The firms have options whether or not to engage in a multinational investment by obtaining a level of profits. As real options apply to manage the risk of inventories related to uncertain demand, retailers and manufacturers often arrange reordering contracts, or call options which permit the retailer to purchase additional merchandise at a pre-decided time for a fixed price and return contracts, or put options which let the retailer to return unsold products at a predetermined salvage value [22]. According to Pandza et al. [23], by delaying investments through waiting for market and technology uncertainty to diminish, the real options logic can identify the value available to firms by waiting before proceeding into a larger commitment.

- *Proposition* 3: Higher uncertainty in market and technology will delay firms' decisions to proceed with additional investment

Overall, the above analysis reveals that the gaps between real world applications and utilizes real options practices to assist managers in deciding the companies' future strategic planning. The propositions facilitate some direction for managers when utilizing option theory.

IMPLICATIONS AND FUTURE RESEARCH DIRECTIONS

From the above discussion, it is evident that a fundamental gap exists between utilizing real option and simulation in real world scenarios. Since it is important for managers to simulate some options before applying the flexibility model, the risk of the uncertainty when the firms invest in any future projects can be alleviated. Several of the articles cited above for flexibility applications such as da Silveria [10], Sawhney [11], Jack and Raturi [9], Pagell and Krause [12], and Duray et al. [14], and with Gamba and Fusari's [18] for simulation under modularity valuing can be expanded to establish some research questions and models. Due to the above research implementation of slightly different flexibility approaches, this section of the present study focuses on Sawhney's [11] article regarding some questions that could be extended when applying real options under certain simulations.

In Table 2, the author studied the upstream (supplier flexibility) and downstream (customer flexibility) and the relationship between input, process, and output flexibility. While not addressing any values to help the firms' generate higher revenue, Sawhney indicates that flexibility will impact the suppliers and customers' flexibility. Therefore, the managers can ask: Which flexibility option can assist the firm to generate the most value through implementation of those decisions? By implementing the flexibility option in each stage, does the firm create incremental value by taking advantage of the upside potential profits? Sawhney specified sixteen propositions for his research; in this study, it is proposed that this research could be modified by using each of his propositions as an option to determine whether to implement or delay the actions under consideration.

Table 2: Sawhney (2006) transformation framework of flexibility

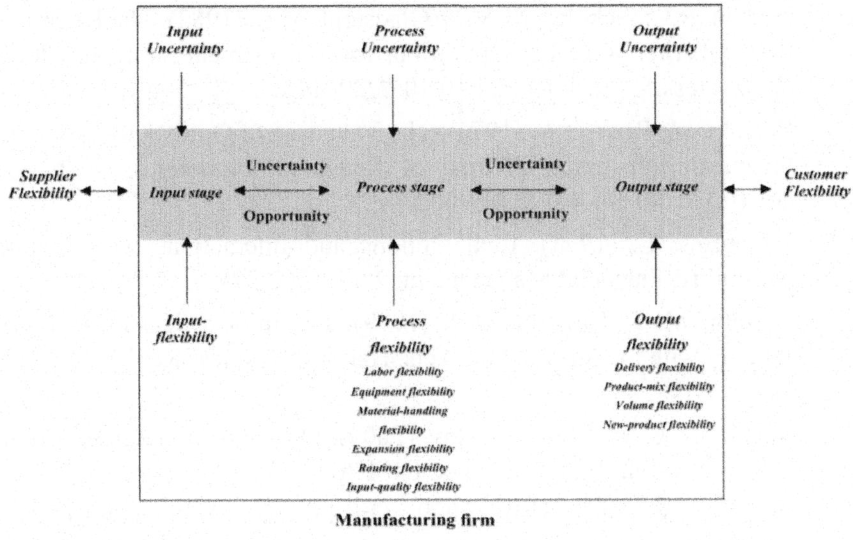

Since Black and Scholes' [2] original option pricing formula was proposed in 1973 to Cox and Ross' [4] risk neutral valuation model, options have been applied from the financial field to various business areas. Corporations have utilized options to help firms alleviate the uncertainty and risk, thereby allowing them to increase value and profits. Many researchers have published related topics in applying flexibility and real options in some of the research. Specifically, JOM includes several articles related to flexibility for their studies and MS focuses on how to implement real options in simulation form. Due to the challenges of institutionalizing the process to connect the real option theory in actual scenarios, few articles have been published on the application side. This review defines the gaps, proposes propositions, and builds on Sawhney's study to arrive at some research questions for the real option decision within a flexibility approach. It is hoped that this proposal can assist managers to make better decisions in generating corporate profits while dealing with the uncertainty and avoiding risk.

REFERENCES

1. Yeo, K.T. & Qiu, F. (2003). The value of management flexibility - a real option approach to investment evaluation. International Journal of Project Management, 21, 243-250.doi:10.1016/S0263-7863(02)00025-X
2. Black, F. & Scholes, M. (1973). The Pricing of Options and Corporate Liabilities. The Journal of Political Economy, 81(3), 637-654. doi:10.1086/260062
3. Merton, R.C., Scholes, M.S., & Gladstein, M.L. (1982). The Returns and Risks of Alternative Put-Option Portfolio Investment Strategies. Journal of Business,55(1), 1-55. doi:10.1086/296153
4. Cox, J.C. & Ross, S.A. (1976). The Valuation of Options for Alternative Stochastic Processes. Journal of Financial Economics, 3, 145-166. doi:10.1016/0304-405X(76)90023-4
5. Trigeorgis, L. (1993). Real Options and Interactions with Financial Flexibility. Financial Management, 22(3), 202-224.
6. Amram, M. & Kulatilaka, N. (1999). Disciplined decisions: aligning strategy with the financial markets. Harvard Business Review, 77(1), 95-104.
7. Boyabatl?, O. and L. B. Toktay. (2004). Operational hedging: a review with discussion. Working Paper. INSEAD.
8. Weiss, D. & Maher, M.W. (2009). Operational hedging against adverse circumstances.Journal of Operations Ma- nagement, 27, 362-373. doi:10.1016/j.jom.2008.10.003

9. Jack, E. P. andRaturi, A. (2002).Sources of volume flexibility and their impact on performance.Journal of Operations Management, 20, 519-548. doi:10.1016/S0272-6963(01)00079-1
10. da Silveira, G. (2006). Effects of simplicity and discipli- ne on operational flexibility: An empirical reexamination of the rigid flexibility model. Journal of Operations Management, 24, 932-947. doi:10.1016/j.jom.2005.11.004
11. Sawhney, R. (2006). Interplay between uncertainty and flexibility across the value-chain: Towards a transformation model of manufacturing flexibility. Journal of Operations Management, 24, 476-493. doi:10.1016/j.jom.2005.11.008
12. Pagell, M. and Krause, D.R. (2004). Re-exploring the relationship between flexibility and the external environment. Journal of Operations Management, 21, 629-649.doi:10.1016/j.jom.2003.11.002
13. Swamidass, P.M., and Newell, W.T. (1987). Manufacturing strategy, environmental uncertainty and performance: a path analytic model. Management Science, 33, 509-524. doi:10.1287/mnsc.33.4.509
14. Duray, R., Ward, P., Milligan, G., & Berry, W. (2000). Approaches to mass customization: configurations and empirical validation. Journal of Operations Management, 18, 605-625.doi:10.1016/S0272-6963(00)00043-7
15. Dilts, D.M. and Pence, K.R. (2006). Impact of role in the decision to fail: An exploratory study of terminated projects. Journal of Operations Management, 24, 378-396. doi:10.1016/j.jom.2004.12.001
16. Folta, T., Delmar, F., & Wennbery, K. (2010). Hybrid Entrepreneurship. Management Science, 56(2), 253-269. doi:10.1287/mnsc.1090.1094
17. Chick, S. & Gans, N. (2009). Economic Analysis of Simulation Selection Problems. Management Science, 55(3), 421-437.doi:10.1287/mnsc.1080.0949
18. Gamba, A. & Fusari, N. (2009). Valuing Modularity as a Real Option. Management Science, 55(11), 1877-1896. doi:10.1287/mnsc.1090.1070
19. Baldwin, C.Y. and Clark, K.B. (2006). The Architectureof Participation: Does Code Architecture Mitigate Free Riding in the Open Source Development Model? Manag-ement Science, 52(7), 1116-1127. doi:10.1287/mnsc.1060.0546
20. Kumar, P. and Turnbull, S.M. (2008). Optimal Patentingand Licensing of Financial Innovations. Management Science, 54(12), 2012-2023. doi:10.1287/mnsc.1080.0898

21. Mello, A.S., Parsons, J.E., and Triantis, A. (1995). An Integrated Model of Multinational Flexibility and Financial Hedging. Journal of International Economics, 39(1-2), 27-51.doi:10.1016/0022-1996(94)01362-V

22. Burnetas, A. and Ritchken, P. (2005). Option Pricing with Downward-Sloping Demand Curves: The Case of Supply Chain Options. Management Science, 51(4), 566-580.doi:10.1287/mnsc.1040.0342

23. Pandza, K., Horsburgh, S., Gorton, K., and Polajnar, A.(2003). A real options approach to managing resources-and capabilities. International Journal of Operations & Production Management, 23(9), 1010-1032. doi:10.1108/01443570310491756

Chapter 10

PERFORMANCE ANALYSIS OF TWO PRIORITY QUEUING SYSTEMS IN TANDEM

Faouzi Kamoun

College of Technological Innovation, Zayed University, Dubai, UAE

ABSTRACT

In this paper, we consider a tandem of two head-of-line (HOL) non-preemptive priority queuing systems, each with a single server and a deterministic service-time. Two classes of traffic are considered, namely high priority and low priority traffic. By means of a generating function approach, we present a technique to derive closed-form expressions for the mean buffer occupancy at each node and mean delay. Finally, we illustrate our solution technique with some numerical examples, whereby we illustrate the starvation impact of the HOL priority scheduling discipline on the performance of the low-priority traffic stream. Our research highlights the important fact that the unfairness of the HOL priority scheduling becomes even more noticeable at the network level. Thus this priority mechanism should be used with caution.

INTRODUCTION

In this contribution, we consider two discrete-time priority queuing systems (hereafter referred to as PQS1 and PQS2) in tandem. Each PQS implements a non-preemptive head-of-line (HOL) priority scheduling with a single server. We consider two types of priority customers, namely high-priority and low priority customers (hereafter referred to as class-1 and class-2, respectively). The number of class-1 and class-2 arrivals within a slot can be correlated random variables.

The main thrust behind our interest to explore the performance of tandem priority queues is that previous studies have focused on an isolated priority queuing system. However, it is well known that in a typical (store and forward) network environment, customers go through a number of switching nodes before reaching their final destinations. For instance, customers belonging to a virtual connection are stored and forwarded through a number of queues in tandem. Similarly, in many manufacturing and service applications (including health-care and banking), customers (or jobs) need to go through a number of prioritized service facilities before completing a single transaction. At the same time, none of the previous studies on tandem queues have considered multi-class priority scheduling schemes. It is well known however that next generation telecommunications networks are being built around the inspiration of having a single costeffective packet-based network that is capable of supporting diverse classes of services, each with its own Quality of Service (QoS) requirement. For instance, realtime traffic is delay and delay-jitter sensitive, while nonreal-time applications are mainly loss sensitive. To achieve this goal, various packet service disciplines that determine the order by which customers are served have been proposed. Among the simplest time-priority scheduling schemes, the non-preemptive head-of-line (HOL) priority scheduling discipline has been proposed to provide differentiated services to the high-priority traffic class. Under this scheme, the server always schedules the delaysensitive traffic first (if present). Many previous studies (see for example [1-4]) have analyzed the performance of discrete-time HOL-priority queuing systems (in terms of system content and/or customer delay) under various assumptions regarding the arrival and service time distributions. In particular, Walraevens et al. [5] analyzed a single server discrete-time HOL priority queuing system in steady-state. The corresponding transient results were reported in [6], while the case of multi-server HOL priority queue is treated in [7]. In [8], Walraevens et al. analyzed the busy periods and the output characteristics of a HOL priority queue. Kamoun [9] analyzed a HOL priority queuing system which is subjected to a Markovian interruption process. However, to our best knowledge, none of the previous contributions addressed the case where customers have to go through more than one priority queuing stage.

The contribution of this paper consists of the practical queuing model being considered, the solution technique being developed to extract explicit expressions for the various performance measures, and the new insights provided on the performance of HOL priority queuing systems in a small "network" environment:

First, we add to the existing literature on non-preemptive priority models by considering two priority-based queuing systems in tandem. This tandem configuration of two PQS stages makes the resulting model more practical and flexible in studying more realistic priority-based queuing systems.

Second, we demonstrate how a generating function can be used to derive closed-form expressions for the mean buffer occupancies and mean delays.

The closedform expressions of these performance measures make it very convenient to understand the effect of various system parameters on the performance of both traffic types. In particular, this research demonstrates how the starvation of the low-priority traffic, due to the HOL priority scheduling discipline, is particularly accentuated at the second stage. Our approach also bypasses the need to characterize or approximate the output process of the first PQS stage. Last, the exact expressions for some key performance measures can also be used as baselines to validate the accuracy of potential approximation techniques.

The remainder of this paper is organized as follows: In Sections 2 and 3, we describe the queuing model and derive an expression for the Joint Probability Generating Function (JPGF) of the system state vector. In Section 4, we derive explicit expressions for some boundary terms appearing in the JPGF. From this JPGF, the marginal PGFs of the system contents at three of the four nodes, as well as several other JPGFs are derived, as illustrated in Section 5. Closed-form expressions for the first moments of the system contents and customer delay are presented in Section 6. Numerical results that provide insights into the behavior of the queuing model are provided in Section 7. In particular, we wanted to further explore the impact of the HOL priority scheduling discipline in a tandem configuration on the performance of class-1 and class-2 customers. Finally, a summary of the main findings of the papers and some suggestions for further research are provided in Section 8.

Analytical Model Description

We consider two discrete-time priority queuing systems (PQS) in a tandem configuration. Each PQS has an infinite storage capacity and a single (FCFS) server. The time axis is divided into equal length slots and customer service completion is synchronized to occur at the slot boundaries. Here a slot is the time period required to transmit exactly one customer from the system. It is assumed that a customer which arrives during a slot cannot be served before the beginning of the next slot. A "conceptual" diagram for our queuing model is depicted in **Figure 1**.

As shown in **Figure 1**, in each PQS, customers are served based on a HOL time-priority scheme, whereby type-1 customers have absolute non-preemptive priority over type-2 customers. Under this priority scheme, the server will always serve type-1 customers (if any). If there are no type-1 customers, then type-2 customers (if any) will be served. The model shown in **Figure 1** can be viewed as two HOL priority queuing systems cascaded in tandem (PQS1 and PQS 2), or as two parallel tandem queuing systems, which are coupled. In the latter case, the first tandem system consist of two high-priority queues (nodes 1 and 3), while the second system consists of the two low-priority queues (nodes 2 and 4). We also note that the input to the high/low priority queues (node 3/4 in PQS 2) consists of new type 1/2 customers joining the network (exogenous arrivals), as well as type 1/2 customers which completed service at the first PQS. It is also clear from the model description that the high-priority traffic stream, traversing nodes 1 and 3, is not interrupted by the HOL priority scheduling mechanism. The low-priority traffic stream, on the other hand, is subjected to two interruptions, at nodes 2 and 4. It is also assumed that inter-stage departures are not allowed.

In the sequel, we denote the number of arrivals of class-j customers to PQS 1 during slot k by $a_{j,k}\,(j=1,2)$. These arrivals constitute a series of i.i.d. random variables with the common joint probability generating function $A(z_1,z_2)=E\left[z_1^{a_{1,k}}z_2^{a_{2,k}}\right]$, independent of k. We define the marginal pgfs of the arrivals to PQS 1 from class-1 and class-2 customers during a slot by $A_1(z_1)=E\left[z_1^{a_{1,k}}\right]$ and $A_2(z_2)=E\left[z_2^{a_{2,k}}\right]$ respectively.

The marginal arrival rates of class-j (j = 1, 2) to PQS 1 are denoted by $\lambda_j=A_j'(1)$. We further denote the total number of arriving customers to PQS 1 during slot kby $a_{1T,k}=a_{1,k}+a_{2,k}$, with corresponding pgf $A_T(z)$. The total arrival rate to PQS 1 is denoted $\lambda_{1T}=A_T'(1)=\lambda_1+\lambda_2$. Similarly, we denote the number of "exogenous" arrivals of class-j customers to PQS 2 during slot k by $b_{j,k}\,(j=1,2)$. These arrivals constitute a series of i.i.d. random variables with the common joint probability generating function $B(z_3,z_4)=E\left[z_3^{b_{1,k}}z_4^{b_{2,k}}\right]$, independent of k. We define the marginal pgfs of the exogenous arrivals to PQS 2 from class-1 and class-2 customers during a slot by $B_3(z_3)=E\left[z_3^{b_{1,k}}\right]$ and $B_4(z_4)=E\left[z_4^{b_{2,k}}\right]$ respectively.

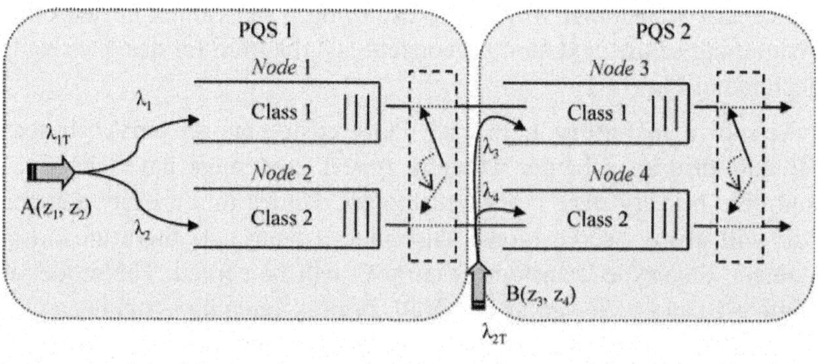

Figure 1: Conceptual diagram of the tandem queuing model with HOL priority.

The marginal exogenous arrival rates of class-j (j = 1, 2) to PQS 2 are denoted by $\lambda_j = B_j'(1)\,(j=3,4)$. We further denote the total number of exogenous arrivals to PQS 2 during slot k by $b_{2T,k} = b_{1,k} + b_{2,k}$, with corresponding pgf $B_T(z)$. The total arrival rate to PQS 2 is denoted by $\lambda_{2T} = B_T'(1) = \lambda_3 + \lambda_4$. The total arrival rate to the whole system is denoted by $\lambda_T = A_T'(1) + B_T'(1) = \lambda_1 + \lambda_2 + \lambda_3 + \lambda_4$.

In the sequel, we assume that the system is stable and that a stochastic equilibrium exists.

Our model allows for the number of class-1 and class-2 arrivals to each PQS, within a slot, to be correlated random variables. This type of correlation can occur for instance when an incoming job (customer) can be split into two parts, an urgent and a non-urgent part, where the urgent part is of class-1, while the non-urgent part is of class-2. Taking into account this correlation in the arrival process leads to more accurate performance results [10]. In the next sections, we will demonstrate that an analysis based on probability generating functions approach can be used to derive some relevant performance measures for the queuing system under consideration.

MATHEMATICAL MODEL

The queuing model, shown in **Figure 1**, can be formulated as a discrete-time multidimensional Markov chain. The state of the system is defined by $(u_{1,k}, u_{2,k}, u_{3,k}, u_{4,k})$ where $u_{i,k}$ is the system content of node i (i = 1 4) at the end of the kth slot. The evolution of the system contents is described by the following system equations:

$$u_{1,k+1} = \left[u_{1,k} - 1\right]^{+} + a_{1,k+1}$$

$$u_{2,k+1} = \begin{cases} \left[u_{2,k} - 1\right]^{+} + a_{2,k+1} & \text{if } u_{1,k} = 0 \\ u_{2,k} + a_{2,k+1} & \text{if } u_{1,k} > 0 \end{cases}$$

$$u_{3,k+1} = \left[u_{3,k} - 1\right]^{+} + b_{1,k+1} + \left[u_{1,k}\right]^{+}$$

$$u_{4,k+1} = \begin{cases} \left[u_{4,k} - 1\right]^{+} + b_{2,k+1} + 1\big|_{u_{2,k}>0 \wedge u_{1,k}=0} & \text{if } u_{3,k} = 0 \\ u_{4,k} + b_{2,k+1} + b_{2,k+1} + 1\big|_{u_{2,k}>0 \wedge u_{1,k}=0} & \text{if } u_{3,k} > 0 \end{cases} \quad (1)$$

where $[x]^{'}$ is a binary-valued random variable which takes the value 1 if x > 0 and 0 otherwise, while $1|_x$ is the indicator function which takes the value 1 if x is true and 0 otherwise.

The first two expressions in (1) follow from the fact that high priority customers are not in influenced by low-priority customers, while a low priority customer can only be served, if there are no high priority customers in PQS1 (which leads to the second equation). The last two expressions in (1) follow from similar arguments, in addition to taking into account the fact that the output of high/low priority queue in PQS1 (if any) will also constitute an input to the high/low priority queue in PQS2. We also took into account the fact that the lowpriority node (2) can only forward a customer to the downstream low-priority node (3) if it is not empty and if the there are no high priority customers in PQS1.

Next, define the JPGF of system contents at the end of the kth slot as follows:

$$Q_k(z_1,z_2,z_3,z_4) = E\left[z_1^{u_{1,k}} z_2^{u_{2,k}} z_3^{u_{3,k}} z_4^{u_{4,k}}\right]$$

$$= \sum_{i=0}^{\infty}\sum_{j=0}^{\infty}\sum_{k=0}^{\infty}\sum_{l=0}^{\infty} z_1^i z_2^j z_3^k z_4^l p_k(i,j,k,l) \tag{2}$$

where

$$p_k(i,j,k,l) = pr\left[u_{1,k}=i, u_{2,k}=j, u_{3,k}=k, u_{4,k}=l\right]$$

is the joint distribution of the system state vector at the end of the kth slot.

It follows that:

$$Q_{k+1}(z_1,z_2,z_3,z_4) = E\left[z_1^{u_{1,k+1}} z_2^{u_{2,k+1}} z_3^{u_{3,k+1}} z_4^{u_{4,k+1}}\right].$$

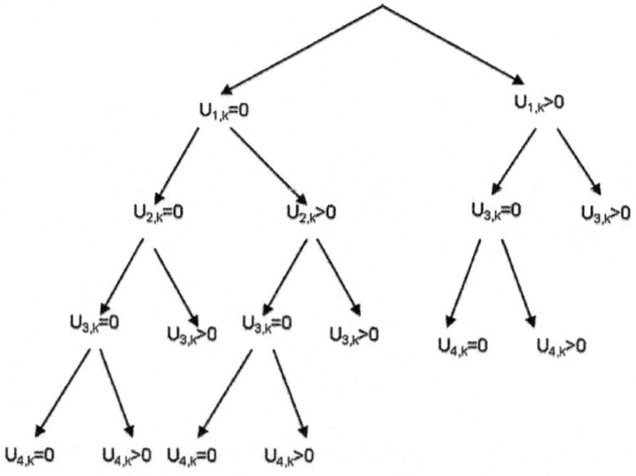

Figure 2: The nine cases used in the derivation of Equation (3).

Taking the z-transform of the system Equations (1), and by considering the nine cases depicted in**Figure 2**, we derive, by means of some standard z-transform manipulations, the following relationship between

$Q_{k+1}(z_1,z_2,z_3,z_4)$ and $Q_k(z_1,z_2,z_3,z_4)$, as illustrated in Equation (3) (below).

Next, we define the steady-state JPGF of the system contents:

$$Q(z)=Q(z_1,z_2,z_3,z_4)\triangleq \lim_{k\to\infty} Q_k(z_1,z_2,z_3,z_4)$$

Applying the above limit to Equation (3) yields (see Equation (4) below):

Although it was not possible to determine all the seven boundary functions appearing in (4), as one of the 2 unknown boundary terms ($Q(0,0,z_3,z_4)$ or $Q(0,z_2,0,z_4)$) could not be determined, we were able to derive the relevant performance measures related to mean system contents at each node and mean delays. It should be noted that the strong coupling between node 4 and the remaining nodes made the direct application of standard pgf techniques to extract all the boundary terms, appearing in (4), an intricate task.

DETERMINATION OF RELEVANT BOUNDARY TERMS

In this section, we show how to progressively determine some unknown boundary terms in order to extract relevant performance measures for the queuing model under consideration.

Determination of Q (z_1, z_2, 0, 0), Q (0, z_2, 0, 0) and Q (0,0,0,0)

First, to determine $Q(z_1,z_2,0,0)$, let $k\to\infty, z_4\to 0$ and then $z_3\to 0$ in Equation (3). Then:

$$Q(z_1,z_2,0,0)=A(z_1,z_2)B(0,0)\cdot\mathrm{K} \tag{5}$$

where:

$$\mathrm{K}=Q(0,0,0,0)+\left.\frac{\partial Q(0,0,0,z_4)}{\partial z_4}\right|_{z_4=0}+\left.\frac{\partial Q(0,0,z_3,0)}{\partial z_3}\right|_{z_3=0} \tag{6}$$

Next, to determine the constant K, we set z_1=z_2=1 in Equation (5), which yields:

$$\mathrm{K}=\frac{Q(1,1,0,0)}{B(0,0)} \tag{7}$$

The boundary constant Q(1,1,0,0) is readily obtained from the normalization condition Q (1,1,1,1) = 1. In fact by setting $z_1 = z_2 = z_3 = z_4 = z$ in Equation (4), it is easy to derive the following expression for the PGF of the total system content:

$$Q_T(z) = \frac{A_T(z)B_T(z)(z-1)Q(z,z,0,0)}{z - A_T(z)B_T(z)} \tag{8}$$

The normalization condition $Q_T(1)=1$ leads to $Q(1,1,0,0)=1-\lambda_T$. By substituting this result back into Equation (7) and then into Equation (5), we immediately obtain the first unknown boundary function:

$$Q(z_1,z_2,0,0)=(1-\lambda_T)A(z_1,z_2) \tag{9}$$

$$\begin{aligned}
&Q_{k+1}(z_1,z_2,z_3,z_4)=E\left[z_1^{n_{1,k+1}} z_2^{n_{2,k+1}} z3^{n_{3,k+1}} z_4^{n_{4,k+1}}\right]\\
&=A(z_1,z_2)B(z_3,z_4)\Big\{Q_k(0,0,0,0)+\frac{1}{z_4}\left[Q_k(0,0,0,z_4)-Q_k(0,0,0,0)\right]+\frac{1}{z_3}\left[Q_k(0,0,z_3,z_4)-Q_k(0,0,0,z_4)\right]\\
&+\frac{z_4}{z_2}\left[Q_k(0,z_2,0,0)-Q_k(0,0,0,0)\right]+\frac{1}{z_2}\left[Q_k(0,z_2,0,z_4)-Q_k(0,0,0,z_4)-Q_k(0,z_2,0,0)+Q_k(0,0,0,0)\right]\\
&+\frac{z_4}{z_2z_3}\left[Q_k(0,z_2,z_3,z_4)-Q_k(0,0,z_3,z_4)-Q_k(0,z_2,0,z_4)+Q_k(0,0,0,z_4)\right]\\
&+\frac{1}{z_1}\left[Q_k(z_1,z_2,z_3,z_4)-Q_k(0,z_2,z_3,z_4)-Q_k(z_1,z_2,0,z_4)+Q_k(0,z_2,0,z_4)\right]\\
&+\frac{z_3}{z_1}\left[Q_k(z_1,z_2,0,0)-Q_k(0,z_2,0,0)\right]+\frac{z_3}{z_1z_4}\left[Q_k(z_1,z_2,0,z_4)-Q_k(0,z_2,0,z_4)-Q_k(z_1,z_2,0,0)+Q_k(0,z_2,0,0)\right]\Big\}
\end{aligned} \tag{3}$$

$$Q(z)=A(z_1,z_2)B(z_3,z_4)\cdot\frac{\begin{aligned}&\{z_4(z_1z_4-z_2z_3)Q(0,z_2,z_3,z_4)+z_2z_3(z_3-z_4)Q(z_1,z_2,0,z_4)\\&+(z_3-z_4)(z_1z_4-z_2z_3)Q(0,z_2,0,z_4)+z_2z_3^2(z_4-1)Q(z_1,z_2,0,0)\\&+z_1(z_3-z_4)(z_2-z_4)Q(0,0,0,z_4)+z_1z_4(z_2-z_4)Q(0,0,z_3,z_4)\\&+z_3(z_4-1)(z_1z_4-z_2z_3)Q(0,z_2,0,0)+z_1z_3(z_4-1)(z_2-z_4)Q(0,0,0,0)\}\end{aligned}}{z_2z_3z_4\left[z_1-A(z_1,z_2)B(z_3,z_4)\right]} \tag{4}$$

The other two related boundary terms are readily derived from Equation (9):

$$Q(0,z_2,0,0)=(1-\lambda_T)A(0,z_2) \tag{10}$$

$$Q(0,0,0,0)=(1-\lambda_T)A_T(0) \tag{11}$$

and therefore for the system to be stable, we require that $Q(0,0,0,0)>0$ or equivalently, $\lambda_T<1$.

Relationship between Q (z_1, z_2, 0, z_4) and Q (0, z_2, 0, z_4)

In this subsection, we determine $Q(z_1,z_2,0,z_4)$ up to the boundary term $Q(0,z_2,0,z_4)$. For this purpose, let $k\to\infty$ and $z_3\to 0$ in Equation (3). Then:

$$Q(z_1,z_2,0,z_4)=A(z_1,z_2)B(0,z_4)\Phi(z_2,z_4) \tag{12}$$

where:

$$\begin{aligned}\Phi(z_2,z_4)&\\ =Q(0,0,0,0)&+\frac{1}{z_4}\left[Q(0,0,0,z_4)-Q(0,0,0,0)\right]\\ &+\left.\frac{\partial Q(0,0,z_3,z_4)}{\partial z_3}\right|_{z_3=0}+\frac{z_4}{z_2}\left[Q(0,z_2,0,0)-Q(0,0,0,0)\right]\\ &+\frac{z_4}{z_2}\left[\left.\frac{\partial Q(0,z_2,z_3,z_4)}{\partial z_3}\right|_{z_3=0}-\left.\frac{\partial Q(0,0,z_3,z_4)}{\partial z_3}\right|_{z_3=0}\right]\\ &+\frac{1}{z_2}\left[Q(0,z_2,0,z_4)-Q(0,0,0,z_4)-Q(0,z_2,0,0)\right.\\ &\qquad\left.+Q(0,0,0,0)\right]\end{aligned}$$

is independent of z_1. Substituting $z_1 = 0$ in Equation (12), we get:

$$\Phi(z_2,z_4)=\frac{Q(0,z_2,0,z_4)}{A(0,z_2)B(0,z_4)}$$ and therefore:

$$Q(z_1,z_2,0,z_4)=\frac{A(z_1,z_2)}{A(0,z_2)}Q(0,z_2,0,z_4) \tag{13}$$

Thus $Q(z_1,z_2,0,z_4)$ is determined up to the boundary term $Q(0,z_2,0,z_4)$.

Determination of Q (0, 0, z, z) and Q (0, z, 0, z)

To determine the boundary term $Q(0,0,z,z)$, we proceed as follows:

First by setting $z_1 = z_2 = x$ and $z_3 = z_4 = z$ in Equation (4), we obtain (see Equation (14) below):

From (9), we note that $Q(x,x,0,0)=(1-\lambda_T)A_T(x)$. Also since $Q(x,x,z,z)$ is analytic inside the polydisk ($(|x|<1,|z|<1)$), then the numerator must be zero whenever the denominator is zero. It follows that:

$$Q(0,0,z,z)$$
$$=(1-\lambda_T)(z-1)\left[\frac{zA_T(Y_z)+(Y_z-z)A_T(0)}{z-Y_z}\right] \tag{15}$$

where Y_z is the unique solution for x in the unit disk $(|z|<1)$ of the equation $x-A_T(x)B_T(z)=0$.

Next, we determine $Q(0,z,0,z)$ by proceedings as follows:

First by setting $z_1 = z_3 = x$ and $z_2 = z_4 = z$ in Equation (4), we obtain (see Equation (16) below):

From (9), we note that $Q(x,x,0,0)=(1-\lambda_T)A_T(x)$. Also from (13):

$$Q(x,z,0,z)=\frac{A(x,z)}{A(0,z)}Q(0,z,0,z) \tag{17}$$

Substituting for $Q(x,x,0,0)$ and $Q(x,z,0,z)$ in (16) and since $Q(x,z,x,z)$ is analytic inside the polydisk $(|x|<1,|z|<1)$, then from Rouche's theorem, we get:

$$Q(0,z,0,z)=(1-\lambda_T)(z-1)A(0,z)\frac{R_z}{z-R_z} \tag{18}$$

where R_z is the unique solution for x of the equation $x=A(x,z)B(x,z)$ in the unit circle $|x|<1$.

JOINT AND MARGINAL PGFS

Having determined some relevant boundary terms, we can now easily derive the marginal PGFs of the system contents at nodes 1, 2, and 3, as well as several other JPGFs. This is illustrated below.

Joint and Marginal PGFs of High-Priority (Class-1) System Contents

Let $P_{13}(z_1,z_3)=Q(z_1,1,z_3,1)$ denote the JPGF of the

$$Q(x,x,z,z)=A_T(x)B_T(z)\cdot\left[\frac{(x-z)Q(0,0,z,z)+z(z-1)Q(x,x,0,0)+(z-1)(x-z)Q(0,0,0,0)}{z[x-A_T(x)B_T(z)]}\right] \tag{14}$$

$$Q(x,z,x,z)=A(x,z)B(x,z)\cdot\left[\frac{(x-z)Q(x,z,0,z)+x(z-1)Q(x,z,0,0)}{z[x-A(x,z)B(x,z)]}\right] \tag{16}$$

system contents of the high-priority queues at nodes 1 and 3. By setting $z_2 = z_4 = 1$ in Equation (4), we get (see Equation (19) below):

To determine the boundary terms $P_{13}(z_1,0)$ and $P_{13}(0,0)$, appearing in (19), we first set $z_2 = z_4 = 1$ in (13):

$$P_{13}(z_1,0)=\frac{A_1(z_1)}{A_1(0)}P_{13}(0,0) \tag{20}$$

Next, from (19), the normalization condition P_{13} (1, 1) = 1 yields $P_{13}(1,0)=1-\lambda_1-\lambda_3$. Therefore setting $z_1=1$ in (20), we get:

$$P_{13}(0,0)=(1-\lambda_1-\lambda_3)A_1(0)$$

$$P_{13}(z_1,0)=(1-\lambda_1-\lambda_3)A_1(z_1) \tag{21}$$

The remaining boundary term $P_{13}(0,z_3)$ is readily obtained by substituting (20) and (21) into (19) and by invoking the analytical property of $P_{13}(0,z_3)$ inside the polydisk $(|z_1|<1;|z_3|<1)$. It follows from Rouche's theorem that:

$$P_{13}(0,z_3)$$
$$=(1-\lambda_1-\lambda_3)(z_3-1)\cdot\frac{z_3A_1(X_{z_3})+(X_{z_3}-z_3)A_1(0)}{z_3-X_{z_3}} \tag{22}$$

where for a given $z_3(|z_3|<1)$, X_{z_3} is the unique solution for z_1 of the equation $z_1=A_1(z_1)B_3(z_3)$ in the unit circle $|z_3|<1$.

Substituting (21) and (22) back into (19), we obtain after few algebraic manipulations:

$$P_{13}(z_1,z_3)$$
$$=(1-\lambda_1-\lambda_3)A_1(z_1)B_3(z_3)\frac{(z_3-1)}{z_3-X_{z_3}}$$
$$\cdot\frac{\left[(z_3-X_{z_3})A_1(z_1)+(z_1-z_3)A_1(X_{z_3})\right]}{z_1-A_1(z_1)B_3(z_3)} \tag{23}$$

From the above, we can obtain expressions for the following:

- Marginal PGF of the number of class-1 customers in queue 1:

$$P_1(z_1)=P_{13}(z_1,1)=(1-\lambda_1)\frac{(z_1-1)A_1(z_1)}{z_1-A_1(z_1)} \tag{24}$$

- Marginal PGF of the number of class-1 customers in queue 3:

$$P_3(z_3) = P_{13}(1, z_3)$$
$$= (1-\lambda_1-\lambda_3)B_3(z_3)$$
$$\cdot \frac{(z_3-1)}{z_3 - X_{z_3}} \frac{\left[z_3 - X_{z_3} + (1-z_3)A_1(X_{z_3})\right]}{1-B_3(z_3)} \quad (25)$$

- PGF of the total number of class-1 customers in the system:

$$P_{13,T}(z) = P_{13}(z,z) = \frac{(1-\lambda_1-\lambda_3)(z-1)A_1(z)^2 B_3(z)}{z - A_1(z)B_3(z)} \quad (26)$$

It should be noted that since the high-priority traffic (traversing nodes 1 and 3) is not influenced by the low-priority traffic (traversing nodes 2 and 4), then the above results for the JPGF $P_{1,3}(z_1, z_3)$ could have been derived by analyzing the performance of the two tandem queues 1 and 3, independently from the rest of the system. We carried this analysis and confirmed that the resulting JPGF of the system contents matches the one derived in Equation (23).

Joint PGFs of the System Contents in PQS1

Let $P_{12}(z_1, z_2) = Q(z_1, z_2, 1, 1)$ denote the JPGF of the system contents of the first priority queuing system (PQS 1). Then from Equation (4):

$$P_{12}(z_1, z_2)$$
$$= A(z_1, z_2) \cdot \frac{(z_1 - z_2)P_{12}(0, z_2) + z_1(z_2 - 1)P_{12}(0,0)}{z_2\left[z_1 - A(z_1, z_2)\right]} \quad (27)$$

The boundary constant $P_{12}(0,0)$ is readily obtained from the normalization condition $P_{12}(1,1) = 1$, yielding $P_{12}(0,0) = 1 - \lambda_1 - \lambda_2$. The boundary term $P_{12}(0, z_2)$ is readily obtained from (27) by applying Rouche's theorem:

$$P_{12}(0, z_2) = (1-\lambda_1-\lambda_2)W_{z_2} \frac{z_2 - 1}{z_2 - W_{z_2}} \quad (28)$$

where for a given $z_2 \left(|z_2| < 1\right)$, W_{z_2} is the unique solution for z_1 of the equation $z_1 = A(z_1, z_2)$ in the unit circle $|z_2| < 1$.

Next, by substituting for $P_{12}(0, z_2)$ and $P_{12}(0,0)$ back into (27), we get:

$$P_{12}(z_1, z_2)$$
$$= (1-\lambda_1-\lambda_2)A(z_1, z_2) \cdot \frac{z_2 - 1}{z_2 - W_{z_2}} \frac{z_1 - W_{z_2}}{z_1 - A(z_1, z_2)} \tag{29}$$

$$P_{13}(z_1, z_3) = A_1(z_1)B_3(z_3) \cdot \left[\frac{z_3(z_3-1)P_{13}(z_1,0) + (z_1 - z_3)P_{13}(0,z_3) + (z_3-1)(z_1-z_3)P_{13}(0,0)}{z_3[z_1 - A_1(z_1)B_3(z_3)]} \right] \tag{19}$$

From the above, we can obtain expressions for the following:

- Marginal PGF of the number of class-1 customers in queue 1:

$$P_1(z_1) = P_{12}(z_1, 1) = (1-\lambda_1)\frac{(z_1-1)A_1(z_1)}{z_1 - A_1(z_1)} \tag{30}$$

- Marginal PGF of the number of class-2 customers in queue 2:

$$P_2(z_2) = P_{1,2}(1, z_2)$$
$$= (1-\lambda_1-\lambda_2)A_2(z_2)\frac{z_2 - 1}{z_2 - W_{z_2}} \frac{1 - W_{z_2}}{1 - A_2(z_2)} \tag{31}$$

- PGF of the total number of class-1 customers in PQS1

$$P_{12,T}(z) = P_{12}(z, z) = \frac{(1-\lambda_1-\lambda_2)(z-1)A_T(z)}{z - A_T(z)} \tag{32}$$

We also observe that the JPGF of the system contents of PQS 1 could have also been be derived by considering PQS 1 as an isolated priority queuing system. Such an analysis of a single stage PQS has been presented in [5] and our expressions (31-32) match those derived therein.

PGFs of Total Number of Customers in PQS2

Let $P_{34,T}(z) = P_{34}(z, z) = Q(1,1,z,z)$ denote the JPGF of the total number if customers in the second priority queuing system (PQS 2). Then from (4):

$$P_{34,T}(z)$$
$$= (z-1)B_T(z)$$
$$\cdot \frac{zQ(1,1,0,0) - Q(0,0,z,z) - (z-1)Q(0,0,0,0)}{z[1 - B_T(z)]} \tag{33}$$

Using (9), (11) and (15), we can determine the unknown boundary terms appearing in the above equation, yielding:

$$P_{34,T}(z) = \frac{(1-\lambda_T)(z-1)B_T(z)\left[z-Y_z-(z-1)A_T(Y_z)\right]}{(z-Y_z)(1-B_T(z))} \tag{34}$$

where, as defined earlier, for a given $z(|z|<1)$, Y_z is the unique solution for x of the equation $x-A_T(x)B_T(z)=0$ in the unit circle $|x|<1$.

PGF of the Total Number of Class-2 Customers in the System

Let $P_{24,T}(z)=P_{24}(z,z)=Q(1,z,1,z)$ denote the PGF of the total number of class-2 customers in the system. Then from Equation (4):

$$P_{24,T}(z) = (z-1)A_2(z)B_4(z)\frac{(1-\lambda_T)A_2(z)-Q(1,z,0,z)}{z\left[1-A_2(z)B_4(z)\right]} \tag{35}$$

The unknown boundary term $Q(1,z,0,z)$ can be determined from (13) and (18), which yields:

$$P_{24,T}(z)=\frac{(1-\lambda_T)(z-1)A_2(z)^2 B_4(z)}{1-A_2(z)B_4(z)}\cdot\frac{1-R_z}{z-R_z} \tag{36}$$

where R_z is the unique solution for x of the equation $x=A(x,z)B(x,z)$ inside the unit circle $|x|<1$.

PGF of the Total System Contents

Let $P_T(z)=Q(z,z,z,z)$ denote the PGF of the total system contents. Then using (4) and (9) we get:

$$P_T(z)=(1-\lambda_T)(z-1)\frac{A_T(z)^2 B_T(z)}{z-A_T(z)B_T(z)} \tag{37}$$

CALCULATION OF THE MOMENTS

To compute the various moments of system contents, we introduce and recall the following notations:

$$\lambda_{ij}=\begin{cases}\dfrac{\partial^2 A(z_1,z_2)}{\partial z_i \partial z_j}\Big|_{z_1=z_2=1} & (i,j)\in\{1,2\}\times\{1,2\}\\[2ex] \dfrac{\partial^2 B(z_3,z_4)}{\partial z_i \partial z_j}\Big|_{z_3=z_4=1} & (i,j)\in\{3,4\}\times\{3,4\}\end{cases}$$

$$\lambda_{1T} = A_T'(1) = \frac{dA_T(z)}{dz}|_{z=1} = \lambda_1 + \lambda_2$$
$$\lambda_{2T} = B_T'(1) = \frac{dB_T(z)}{dz}|_{z=1} = \lambda_3 + \lambda_4$$

$$A_T''(1) = \frac{d^2 A_T(z)}{dz^2}|_{z=1} = \lambda_{11} + \lambda_{22} + 2\lambda_{12}$$
$$B_T''(1) = \frac{d^2 B_T(z)}{dz^2}|_{z=1} = \lambda_{33} + \lambda_{44} + 2\lambda_{34} \tag{38}$$

Next, we observe that the functions $X_{z_3}, W_{z_2}, Y_z, R_z$, appearing in the various pgfs of the system contents, can only be explicitly defined under simple arrival processes. Fortunately, the derivation of the first and higher moments of the system contents involves evaluating these functions and their first and higher order derivatives at $z_i = 1$. These can be readily computed in closed-form.

From the corresponding pgfs (24 - 26, 30 - 32, and 34, 36 - 37), the mean system content of class-j customers (j = 1, 2) and the mean total system contents are readily obtained:

- Mean system content of class-1 customers at node 1:

$$\bar{N}_1 = P_1(1) = \lambda_1 + \frac{\lambda_{11}}{2(1-\lambda_1)} \tag{39}$$

- Mean system content of class-1 customers at node 3:

$$\bar{N}_3 = \lambda_1 - \frac{\lambda_{11}}{2(1-\lambda_1)} + \frac{\lambda_{11} + \lambda_{33} + 2\lambda_3(1-\lambda_3)}{2(1-\lambda_1-\lambda_3)} \tag{40}$$

- Total mean system contents of class-1 customers:

$$\bar{N}_{13,T} = P_{13,T}'(1) = 2\lambda_1 + \lambda_3 + \frac{\lambda_{11} + 2\lambda_1\lambda_3 + \lambda_{33}}{2(1-\lambda_1-\lambda_3)} \tag{41}$$

- Mean system content of class-2 customers at node 2:

$$\bar{N}_2 = P_2'(1) = \lambda_2 + \frac{2\lambda_{12} + \lambda_{22}}{2(1-\lambda_{1T})} + \frac{\lambda_2\lambda_{11}}{2(1-\lambda_{1T})(1-\lambda_1)} \tag{42}$$

- Total mean system contents of class-2 customers:

$$\begin{aligned}&\overline{N}_{24,T}\\&=P_{24,T}'(1)=2\lambda_2+\frac{A_T''(1)+B_T''(1)+2\lambda_{2T}(1-\lambda_{2T})}{2(1-\lambda_T)}\\&\qquad-\frac{2\lambda_3(1-\lambda_3)+\lambda_{11}+\lambda_{33}}{2(1-\lambda_1-\lambda_3)}\end{aligned} \tag{43}$$

- Mean system content of class-2 customers at node 4:

The mean number of class-2 customers at node 4 can be derived by subtracting the mean system content of class-2 customers at node 2 from the total mean system contents of class-2 customer. Hence, using (42) and (43), we get:

$$\begin{aligned}&\overline{N}_4=\overline{N}_{24,T}-\overline{N}_2\\&=\lambda_2+\frac{A_T''(1)+B_T''(1)+2\lambda_{2T}(1-\lambda_{2T})-2\lambda_{12}-\lambda_{22}}{2(1-\lambda_{1T})}\\&\quad-\frac{2\lambda_3(1-\lambda_3)+\lambda_{11}+\lambda_{33}}{2(1-\lambda_1-\lambda_3)}-\frac{\lambda_1\lambda_2}{2(1-\lambda_{1T})(1-\lambda_1)}\end{aligned} \tag{44}$$

- Total mean system contents at PQS1:

$$\overline{N}_{12,T}=P_{12,T}'(1)=\lambda_{1T}+\frac{A_T''(1)}{2[1-\lambda_{1T}]} \tag{45}$$

- Total mean system content at PQS2:

$$\begin{aligned}&\overline{N}_{34,T}=P_{34,T}'(1)\\&=\lambda_T+\frac{A_T''(1)+B_T''(1)+2\lambda_{2T}\lambda_{1T}}{2(1-\lambda_T)}-\frac{A_T''(1)}{2(1-\lambda_{1T})}\end{aligned} \tag{46}$$

- Total mean system content

$$\overline{N}_T=P_T'(1)=\lambda_T+\lambda_{1T}+\frac{A_T''(1)+B_T''(1)+2\lambda_{1T}\lambda_{2T}}{2(1-\lambda_T)} \tag{47}$$

It can be easily verified that Equations (39-47) satisfy the expected results:

$$\begin{aligned}\overline{N}_{12,T}&=\overline{N}_1+\overline{N}_2\\\overline{N}_{34,T}&=\overline{N}_3+\overline{N}_4\\\overline{N}_{13,T}&=\overline{N}_1+\overline{N}_3\\\overline{N}_T&=\overline{N}_1+\overline{N}_2+\overline{N}_3+\overline{N}_4\end{aligned} \tag{48}$$

By using Little's law, the following mean customer delays are readily obtained:

- Mean delay at each node:

$$\bar{d}_i = \frac{\bar{N}_i}{\lambda_i} \tag{49}$$

- Mean delay $\bar{d}_{cj}$ **of an arbitrary class-j customer (j = 1, 2):**

$$\bar{d}_{c1} = \frac{\lambda_1}{\lambda_1 + \lambda_3}\bar{d}_1 + \frac{\lambda_3}{\lambda_1 + \lambda_3}\bar{d}_3 = \frac{\bar{N}_{13,T}}{\lambda_1 + \lambda_3}$$

$$\bar{d}_{c2} = \frac{\lambda_2}{\lambda_2 + \lambda_3}\bar{d}_2 + \frac{\lambda_4}{\lambda_2 + \lambda_4}\bar{d}_4 = \frac{\bar{N}_{24,T}}{\lambda_2 + \lambda_4} \tag{50}$$

- Total average delay $\bar{d}_T$ **in the system:**

$$d_T = \frac{\sum_{i=1}^{4}\bar{N}_i}{\lambda_T} = \frac{\bar{N}_T}{\lambda_T} \tag{51}$$

NUMERICAL EXAMPLES AND DISCUSSIONS

In this section, we illustrate our approach through some numerical examples. In particular, we will further probe into the influence of HOL priority on the performance of low-priority traffic stream. For this purpose, we consider two-dimensional Binomial exogenous arrival processes with JPGFs:

$$A(z_1, z_2) = \left(1 - \frac{\lambda_1}{N}(1 - z_1) - \frac{\lambda_2}{N}(1 - z_2)\right)^N$$

$$B(z_3, z_4) = \left(1 - \frac{\lambda_3}{N}(1 - z_3) - \frac{\lambda_3}{N}(1 - z_4)\right)^N$$

Further, we define $v_i = \lambda_i/\lambda_T$ (i = 1, 2, 3) as the fraction of arrivals to node i among the total arrivals. For the numerical calculations, we have set N = 16 and $(\upsilon_1, \upsilon_2, \upsilon_3) = (0.45, 0.25, 0.20)$.

In **Figure 3**, the mean customer delay at each node as well as the total average delay, are plotted as a function of the total arrival rate λ_T. The mean delay $\bar{d}_i$ at each node and the total average delay $\bar{d}_T$ were computed by substituting for $\bar{N}_i$ as given in Equations (39), (40), (42), and (44) into (49) and (51), respectively.

Figure 4 shows the mean delay of an arbitrary class-j customer (j = 1, 2) and the total average delay as a function of the total arrival rate λ_T. In this case, the mean delay λ_T for each class j customer was computed by substituting for $\bar{N}_{13,T}$ and $\bar{N}_{24,T}$, as given in Equations (41) and (43), into (50).

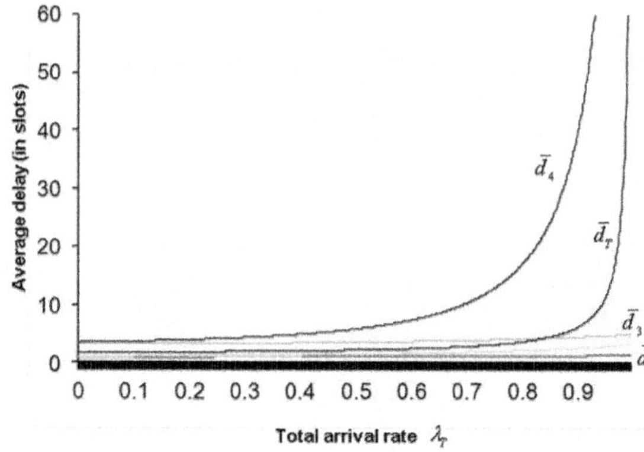

Figure 3: Mean delays versus total arrival rate λ_T.

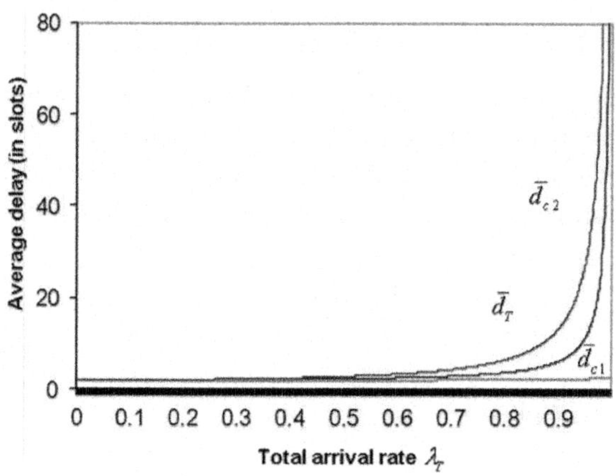

Figure 4: Mean delay of a class-j customer (j = 1, 2).

As may be seen from the above figures, the HOL priority scheduling has a significant influence on the mean delay of the low-priority (class-2) customers. This starvation effect is even more significant at node 4. To this regards, recall that class-2 customers arriving at the access node (2) are subjected to two consecutive HOL interruption processes due to the potential presence of higher priority customers at nodes 1 and 3. The above results not only echo but also amplify previous concerns related to the unfairness of the static HOL priority scheduling mechanism for low-priority traffic. Therefore HOL scheduling should be used with caution.

Our research also suggests that more research is needed to come-up with efficient priority mechanisms that exhibit better fairness and that can be easily implemented in real-time. Earlier contributions in this regards focused on a class of dynamic priority schemes that are based on the queue-length-threshold scheduling discipline. Under this scheme, the priority scheduling can dynamically change, based on the queue lengths (see for example [11-15]) [16]. Some More recent contributions include the work of Maertens et al. [17-19] on priority jumping mechanisms and the contribution of De Vuyst et al. [20,21] on reservation-based priority disciplines.

CONCLUSIONS AND FUTURE RESEARCH

In this paper, we presented an exact analysis of a two HOL priority queuing systems in a tandem configuration. For each PQS, our model allows for possible correlation between the numbers of arrivals of the two classes of customers during a slot. We showed how a generating function approach can be used to derive closed-form expressions for several performance measures. Finally we have demonstrated, via numerical examples, the negative impact the HOL priority scheduling on performance of the low-priority traffic.

This work can be further explored in many directions. For instance, it would be interesting to derive expressions for the pgf of customer delay and characterize the endto-end delay of a tagged customer. The characterization of the asymptotic behavior of the tail distributions of the various system contents is still an open research issue.

Finally, our analysis can also be extended to handle the case where customers leaving the first PQS stage are allowed to leave the system with a pre-assigned probability. These are left for future research.

REFERENCES

1. A. Khamisy and M. Sidi, "Discrete-Time Priority Queues with Two-State Markov Modulated Arrivals," Stochastic Models, Vol. 8, No. 2, 1992, pp. 337-357.doi:10.1080/15326349208807228
2. M. Sidiand and A. Segall, "Structured Priority Queuing Systems with Applications to Packet-Radio Networks," Performance Evaluation, Vol. 3, No. 4, 1983, pp. 265- 275.doi:10.1016/0166-5316(83)90036-6
3. T. Takine, B. Sengupta and T. Hasegawa, "An Analysis of a Discrete-Time Queue for Broadband ISDNwith Priorities among Traffic Classes," IEEE Transactions on Communications, Vol. 42, No. 2-4, 1994, pp. 1837-1845.doi:10.1109/TCOMM.1994.582893

4. T. Takine, "A NonpreemptivePriority MAP/G/1 Queue with Two Classes of Customers," Journal of Operations Research Society of Japan, Vol. 39, No. 2, 1996, pp. 266- 290.
5. J. Walraevens, B. Steyaertand H. Bruneel," Performance Analysis of a Single-Server ATM Queue with a Priority Scheduling," Computers and Operations Research, Vol. 30, No. 12, 2003, pp. 1807-1829. doi:10.1016/S0305-0548(02)00108-9
6. J. Walraevens, D. Fiems and H. Bruneel, "Transient Analysis of a Discrete-Time Priority Queue," Proceedings of the 12th International Conference on Analytical and Stochastic Modelling Techniques and Applications (ASMTA 2005), Riga, June 2005, pp. 17-24.
7. K. Laevens and H. Bruneel, "Discrete-Time Multiserver Queues with Priorities," Performance Evaluation, Vol. 33, No. 4, 1998, pp. 249-275. doi:10.1016/S0166-5316(98)00024-8
8. J. Walraevens, D. Fiems, S. Wittevrongel and H. Bruneel, "Calculation of Output Characteristics of a Priority Queue through a Busy Period Analysis," European Journal of Operational Research, Vol. 198, No. 3, 2009, pp. 891-898.doi:10.1016/j.ejor.2008.11.018
9. F. Kamoun, "Performance Analysis of a Non-Preemptive Priority Queuing System Subjected to a Correlated Markovian Interruption Process," Computers and Operations Research, Vol. 35, No. 12, 2008, pp. 3969-3988. doi:10.1016/j.cor.2007.06.001
10. J. Walraevens, D. Fiems and H. Bruneel, "The DiscreteTime Preemptive Repeat Identical Priority Queue," Queuing Systems, Vol. 53, No. 4, 2006, pp. 231-243.doi:10.1007/s11134-006-7770-x
11. R. Chipalkatti, J. F. Kurose and D. Towsley, "Scheduling Policies for Real-Time and Nonreal-Time Traffic Packet Switching Node," Proceedings of the IEEE INFOCOM '89, Ottawa, April 1998, pp. 774-783.
12. B. I. Choi and Y. Lee, "Performance Analysis of a Dynamic Priority Queue for Traffic Control of Bursty Traffics in ATM Networks," IEEE Proceedings of Communications, Vol. 148, No. 3, 2001, pp. 181-187. doi:10.1049/ip-com:20010115
13. D. I. Choi, B. D. Choi and D. Sung, "Performance Analysis of Priority Leaky Bucket Scheme with QueueLength-Threshold Scheduling Policy," IEEE Proceedings of Communications, Vol. 145, No. 6, 1998, pp. 395-401. doi:10.1049/ip-com:19982287
14. C. Knessl, D. I. Choi and C. Trier, "A Dynamic Priority Queue Model for Simultaneous Service of Voice and Data Traffic," SIAM Journal on

Applied Mathematics, Vol. 63, No. 2, 2002, pp. 398-422. doi:10.1137/S0036139901390842

15. J. T. Lee and Y. H. Kim, "Performance Analysis of a Hybrid Priority Control Scheme for Input and Output Queuing ATM Switches," Proceedings of INFOCOM '98, San Francisco, 29 March-2 April 1998, pp.1470-1477. doi:10.1109/INFCOM.1998.662965
16. T. Maertens, J. Walraevens and H. Bruneel, "Performance Analysis of a Single-Server Queue with HOL-PJ Priority Scheduling Discipline," Proceedings of the Second International Working Conference on Performance Modelling and Evaluation of Heterogeneous Networks (HETNETs '04), Ilkley, 26-28 July 2004, pp. 1-10.
17. T. Maertens, J. Walraevens and H. Bruneel, "On Priority Queues with Priority Jumps," Performance Evaluation, Vol. 63, No. 12, 2006, pp. 1235-1252.doi:10.1016/j.peva.2005.12.003
18. T. Maertens, J. Walraevens and H. Bruneel, "A Modified HOL Priority Scheduling Discipline: Performance Analysis," European Journal of Operational Research, Vol. 180, No. 3, 2007, pp. 1168-1185. doi:10.1016/j.ejor.2006.05.004
19. T. Maertens, J. Walraevens, M. Moeneclaey and H. Bruneel," Performance Analysis of a Discrete-Time Queuing System with Priority Jumps," International Journal of Electronics and Communications, Vol. 63, No. 10, 2009, pp. 853-858.doi:10.1016/j.aeue.2008.07.004
20. S. De Vuyst, S. Wittevrongel, D. Fiems and H. Bruneel, "Controlling the Delay Trade-off Between Packet Flows UsingMultiple Reserved Places," Performance Evaluation, Vol. 65, No. 6-7, 2008, pp. 484-511. doi:10.1016/j.peva.2007.12.008
21. S. De Vuyst, D. Wittevrongel and H. Bruneel, "Place Reservation: Delay Analysis of a Novel Scheduling Mechanism," Computers and Operations Research, Vol. 35, No. 8, 2008, pp. 2447-2462. doi:10.1016/j.cor.2006.12.003

Chapter 11

ON A DYNAMIC OPTIMIZATION TECHNIQUE FOR RESOURCE ALLOCATION PROBLEMS IN A PRODUCTION COMPANY

Chuma R. Nwozo[1], Charles I. Nkeki[2]

[1]Department of Mathematics, University of Ibadan, Ibadan, Nigeria

[2]Department of Mathematics, University of Benin, Benin City, Nigeria

ABSTRACT

This paper examines the allocation of resource to different tasks in a production company. The company produces the same kinds of goods and wants to allocate m number of tasks to 50 number of machines. These machines are subject to breakdown. It is expected that the breakdown machines will be repaired and put into operation. From past records, the company estimated the profit the machines will generate from the various tasks at the first stage of the operation. Also, the company estimated the probability of breakdown of the machines for performing each of the tasks. The aim of this paper is to determine the expected maximize profit that will accrue to the company over T horizon. The profit that will accrued to the company was obtained as ₦4,571,100,000 **after 48 weeks of operation. At the** infinity horizon, the profit was obtained to be ₦20,491,000,000. It was found that adequate planning, prompt and effective maintenances can enhance the profitability of the company.

INTRODUCTION

We consider the allocation of tasks to different machines in a production company. A certain number of machines is proposed to be purchased at the beginning of a planning horizon. From statistics, the company has an estimate of the profit each tasks is to yield at the first stage of operation. Also, the company estimates the probability of breakdown of the machines allocated to each tasks. When a machine breaks down, it goes in for repairs after which it returns to the factory for re-allocation at the beginning of the next period.

In this paper, we formulate the problem as a dynamic optimization (DO). Our approach builds on previous research. [1] used the stochastic programming technique of dynamic Programming in financial asset allocation problems for designing low-risk portfolios. [2] proposed the idea of using a parsimonious sufficient static in an application of approximate dynamic programming to inventtory management. [3] described an algorithm for computing parameter values to fit linear and separable concave approximations to the value function for large-scale problems in transportation and logistics. [4] described a more complicated variation of the algorithm that implores execution time and memory requirements. The improvement is critical for practical applications to realistic large-scale problems. [5] used DO for large-scale asset management problems for both single and multiple assets. [6] extended an approximate DO method to optimize the distribution operations of a company manufacturing certain products at multiple production plants and shipping to different customer locations for sales. [7] considered the allocation of buses from a single station to different routes in a transportation company in Nigeria.

In this section, we consider the methodology adopted in this paper. We start with the problem formulation.

PROBLEM FORMULATION

In this section, we consider the methodology adopted in this paper. We start with the problem formulation. Given a certain number of tasks that are to be allocated to different machines at the beginning of each time period, we expect some machine(s) breakdown at the end of each period. Due to the uncertainty in the number of breakdown machine(s), we assume that the states of the machines are random. The company must know the number of machines available for the next period before decision will be made on how to allocate the tasks to the remaining machines. The number of machines to be put into operation in the next period depends on the number of breakdown at the end of the previous period. Our aim is to maximize the total expected profit over a timehorizon. We define the following notations which presented in **Table 1**.

In the next subsection, we define the one-period expected profit function and formulate the problem as a dynamic program.

THE OBJECTIVE AND ONE-PERIOD EXPECTED PROFIT FUNCTION

If the profit for allocating the k task to the machines at period t is Ψ_t^k, the state of the machines is s_t, number of machines allocated to operate on task k at period t under policy π, is $x_t^{\pi k}$ and the number of break down machines is b_t, then the profit that will accrue to the company over T-horizon is given by

$$\sum_{t=0}^{T}\sum_{k=1}^{m}\Psi_t^k\left(x_t^{\pi k}\left(s_{t-1}(b)\right)\right).$$

The expected maximum profit that will accrue to the company under policy π is given by

$$\mathrm{Y}_t^{\pi}(s_t)=E\left[\max_{x_t\in X(s_t)}\sum_{t=0}^{T}\sum_{k=1}^{m}\beta^t\Psi_t^k\left(x_t^{\pi k}\left(s_{t-1}(b)\right)\right)\right] \tag{1}$$

subject to

$$\sum_{K=1}^{m}x_t^{\pi k}\left(s_{t-1}(b)\right)\le s_t,\quad t=0,1,\cdots,T \tag{2}$$

$$x_t^{\pi k}\ge 0, t=1,\cdots,T; k=1,\cdots,m$$

Note: $X(s_t)$ is the set of possible solution of problem (1). Conditioning (1) on $s_t\in S$. We now have the following optimization problem (3)

$$\mathrm{Y}_t^{\pi}(s_t)$$
$$=E\left[\max_{x^{\pi}\in X(s)}\left\{\sum_{t'=t}^{T}\sum_{k=1}^{m}\beta^t\Psi_{t'}^k\left(x_{t'}^{\pi k}\left(s_{t'-1}(b)\right)\right)\right\}\Big/ s_t\in S_t\right]. \tag{3}$$

Problem (1.3) maximizes the expected profit over $X(s_t)$ subject to

$$\sum_{k=1}^{m}x_t^{\pi k}\left(s_{t-1}(b)\right)\le s_t, t=0,1,\cdots,T$$

For the profit function $\Psi:S\to\mathbb{R}$, if we accumulate the profit of the first T-stage and add to it the terminal profit $\Psi_T(s_T)=\sum_{k=1}^{m}\Psi_T^k(s_T)$ then (3) becomes

$$\mathrm{Y}_t^{\pi}(s_t)=E\left[\max_{x^{\pi}\in X(s)}\left\{\sum_{t'=t}^{T-1}\sum_{k=1}^{m}\beta^{t'}\Psi_{t'}^k\left(x_{t'}^{\pi k}\left(s_{t'-1}(b)\right)\right)\right\}\right.$$
$$\left.+\beta^T\Psi_T(s_T)/s_t\in S\right] \tag{4}$$

subject to $\sum_{k=1}^{m}x_{t'}^{\pi k}\left(s_{t'-1}(b)\right)\le s_t, t=0,1,\cdots,T$.

$$x_t^{\pi k}\ge 0, t=1,\cdots,T; k=1,\cdots,m.$$

DYNAMIC PROGRAMMING FORMULATION AND OPTIMALITY

Using s_t as the state variable at period t and S as the state space, we can formulate the problem as a dynamic program. The number of breakdown machines for task k at period t is given by $P_k x_t^k$, where P_k is the probability of break down machines for task k. Hence, total number of break down machines for all the tasks is given by $\sum_{k=1}^{m} P_k x_t^k, t = 1, 2, \cdots, T$.

Table 1: Notations and their Definitions

Notation	Definition
β	discount factor, $0 < \beta < 1$.
S	the state space *i.e.* the set of all machines
T	set of time periods in the planning horizon.
π	is a rule which chooses an action based on current state of the system (policy).
$x_t^{\pi^k}$	number of machines allocated to task k at period t under the policy, π, $t \in [0, T]$.
St	number of functional machines at period t; $s_t \in S$, $t \in [0, T]$
bt	number of breakdown machines at period t, $t \in [0, T]$
$x_t^k(s_{t-1}(b))$	readily available machines to be allocated in the next period.
Ψ_t^k	expected return from task k at period t., $t \in [0, T]$
Π	set of all admissible policy; $\pi \in \Pi$
s_0	number of machines at the beginning of the planning horizon.
$Y_t(s_t)$	objective function of our system, $t \in [0, T]$

We therefore have that

$$b_t = \sum_{k=1}^{m} P_k x_t^k$$

Let s_t be the number of machines to be allocated in period t and let α be the percentage of break down (but repaired) machines that are expected to join the functional ones in period t, then the transformation equation for the system is given by

$$s_t = s_{t-1} - (1-\alpha)\sum_{k=1}^{m} P_k x_t^k, t = 1, \cdots, T \quad (5)$$

Observe that the transformation equation is a random variable.

We now set 1 – α = β in (5) to have,

$$s_t = s_{t-1} - \beta\sum_{k=1}^{m} P_k x_t^k, t = 1, \cdots, T \quad (6)$$

In this case, the optimal policy can be found by computing the value functions through the optimization problem

$$F_t^{\pi}(s_t) = \max_{x^{\pi} \in X(s_t)} \sum_{k=1}^{m} \Psi_t^k \left(x_t^{\pi^k} \left(s_{t-1}(b) \right) \right)$$
$$+ E\{F_t(s_t)\} / S_t = s_t \tag{7}$$

subject to

$$\sum_{k=1}^{m} x_t^{\pi^k} \left(s_{t-1}(b) \right) \le s_t,\ t = 1, 2, \cdots, T.$$

Equivalently,

$$\xi_t + \sum_{k=1}^{m} x_t^{\pi^k} \left(s_{t-1}(b) \right) = s_t, t = 1, \cdots, T; \xi_t \ge 0 \tag{8}$$

$$x_t^{\pi^k} \ge 0, t = 1, \cdots, T; k = 1, \cdots, m.$$

Since all the available functional machines must be allocated in the next period, we have that $\xi_t = 0$, for all $t \in T$, which is the slack variable.

We show that (4) is equivalent to (7), and then use (4) and (7) interchangeably. The theorem below establish this claim.

Lemma 1.1: Let s_t be a state variable that captures the relevant history up to time t, and let $\Upsilon_t(s_{t+1})$ be some function measured at t' ≥ t + 1 conditionalon the random variable s_t. Then, $E\left[E\{\Upsilon_{t'} | s_{t+1}\} | s_t \right] = E\left[\Upsilon_{t'} | s_t \right]$.

For the proof, see [3].

Theorem 1.1: Suppose that $\Upsilon_t^{\pi}(s_t)$ satisfies (4) and $F_t^{\pi}(s_t)$ satisfies (7), then $\Upsilon_t^{\pi}(s_t) = F_t^{\pi}(s_t)$.

Proof: We are to show that $\Upsilon_t^{\pi}(s_t) = F_t^{\pi}(s_t)$. We first use a standard method of DP. Obviously $\Upsilon_T^{\pi}(s_T) = F_T^{\pi}(s_T) = \Psi_T(s_T)$.

Suppose that it hold for t + 1, t + 2, ···, T, then we show that it is true for t. We now write

$$F_t^{\pi}(s_t) = \max_{x^{\pi} \in X(s_t)} \sum_{k=1}^{m} \Psi_t^k \left(x_t^{\pi^k} \left(s_{t-1}(b) \right) \right)$$
$$+ E\left[E \max_{x_t} \sum_{t'=t+1}^{T-1} \sum_{k=1}^{m} \Psi_{t'}^k \left(x_{t'}^{\pi^k} \left(s_{t'}(b) \right) \right) \right.$$
$$\left. + \beta^T \Psi_T(s_T) | S_t = s_t \right]$$

Applying Lemma 1.1, we have

$$F_t^{\pi}(s_t) = \max_{x^{\pi} \in X(s_t)} \sum_{k=1}^{m} \Psi_t^k \left(x_t^{\pi^k} \left(s_{t-1}(b) \right) \right)$$

$$+ E\left[E \max_{x_t} \sum_{t'=t+1}^{T-1} \sum_{k=1}^{m} \Psi_{t'}^k \left(x_{t'}^{\pi^k} \left(s_{t'}(b) \right) \right) \right.$$

$$\left. + \beta^T \Psi_T(s_T) \middle| S_t = s_t \right]$$

When we condition on $x_t^{\pi^k}(s_t(b))$, we obtain

$$F_t^{\pi}(s_t) = E\left[\max_{x^{\pi} \in X(s_t)} \sum_{t'=t}^{T-1} \sum_{k=1}^{m} \Psi_t^k \left(x_t^{\pi^k} \left(s_{t-1}(b) \right) \right) + \Psi_T(s_T) \middle| s_t \right]$$

$$= \Upsilon_t^{\pi}(s_t).$$

For any given objective function, we desire to find the best possible policy, π, that optimizes it, that is, we search for

$$\Upsilon_t^*(s_t) = \max_{\pi \in \Pi} \Upsilon_t^{\pi}(s_t)$$
.

This is obtained by solving the optimality equation

$$F_t(s_t) = \max_{x^{\pi^k} \in X(s_t)} \left[\sum_{k=1}^{m} \Psi_t^k \left(x_t^{\pi^k} \left(s_{t+1}(b) \right) \right) \right.$$

$$\left. + E\{ F_{t+1}(s_t) \} \right]$$

$$= \max_{x^{\pi^k} \in X(s_t)} \left[\sum_{k=1}^{m} \Psi_t^k \left(x_t^{\pi^k} \left(s_{t-1}(b) \right) \right) \right.$$

$$\left. + F_{t+1}\left(s_t - \beta \sum_{k=1}^{m} P_k x_{t+1}^k \right) \right] \qquad (9)$$

If we find the set of F's that solves (9), then we have found the policy that optimizes $\Upsilon_t^{\pi}(s_t)$. The result below establishes this claim.

Theorem 1.2: The expression $F_t^{\pi}(s_t)$ is a solution to equation (1.9) if and only if

$$\Upsilon_t^*(s_t) = F_t(s_t) = \max_{\pi \in \Pi} Y_t^{\pi}(s_t)$$
.

The if part: This is shows by induction that $F_t(s_t) \geq Y_t^*(s_t)$, for all $s_t \in S$ and $t = 0, 1, \cdots, T-1$.

Since $F_T(s_T) = \Psi_T(s_T) = Y_T^{\pi}(s_T)$ for all s_T and $\pi \in \Pi$, we have that $F_T(s_T) = Y_T^*(s_T)$.

Suppose that $F_t(s_t) \geq Y_t^*(s_t) \geq Y_t^{\pi}(s_t)$, for t = n + 1, n + 2, ⋯, T, and let $\pi \in \Pi$ be an arbitrary policy. For t = n, we obtain the optimality equation as follows

$$F_n(s_n) = \max_{x^{\pi^k} \in X(s_n)} \left[\sum_{k=1}^{m} \Psi_n^k \left(x_n^{\pi^k} \left(s_{n-1}(b) \right) \right) + F_{n+1}\left(s_n - \beta \sum_{k=1}^{m} P_k x_{n+1}^k \right) \right].$$

By induction hypothesis, $F_{n+1}(s_{n+1}) \geq Y_{n+1}^*(s_{n+1})$. So we have

$$F_n(s_n) = \max_{x^{\pi^k} \in X(s_n)} \left[\sum_{k=1}^{m} \Psi_n^k \left(x_n^{\pi^k} \left(s_{n-1}(b) \right) \right) + Y_{n+1}^{\pi}\left(s_n - \beta \sum_{k=1}^{m} P_k x_{n+1}^k \right) \right], s' = s_{t+1}.$$

Also, we have that $Y_{n+1}(s_{n+1}) \leq Y_{n+1}^*(s_{n+1})$ for an arbitrary policy, $\pi \in \Pi$, hence

$$\begin{aligned} F_n(s_n) &\geq \max_{x^{\pi^k} \in X(s_n)} \left[\sum_{k=1}^{m} \Psi_n^k \left(x_n^{\pi^k} \left(s_{n-1}(b) \right) \right) + Y_{n+1}^{\pi}\left(s_n - \beta \sum_{k=1}^{m} P_k x_{n+1}^k \right) \right], s' = s_{t+1} \\ &\geq Y_n^{\pi}(s_n), \text{for all } \pi \in \Pi \\ &= \max_{\pi \in \Pi} Y_n^{\pi}(s_n). \end{aligned}$$

Only if part: We show that for any $\in > 0$, there exists $\pi \in \Pi$ that satisfies the following:

$$Y_n^{\pi}(s_n) + (T-n) \in \geq F_n(s_n) \tag{10}$$

We now define $F_n(s_n)$ as follows:

$$F_n(s_n) = \max_{x^{\pi^k} \in X(s_n)} \left[\sum_{k=1}^{m} \Psi_n^k \left(x_n^{\pi^k} \left(s_{n-1}(b) \right) \right) + F_{n+1} \left(s_n - \beta \sum_{k=1}^{m} P_k x_{n+1}^k \right) \right] \tag{11}$$

Let $x_n(s_n)$ be the decision rule that solves (1) and satisfies the following:

$$F_n(s_n) = \sum_{k=1}^{m} \Psi_n^k \left(x_n^{\pi^k} \left(s_{n-1}(b) \right) \right) + F_{n+1} \left(s_n - \beta \sum_{k=1}^{m} P_k x_{n+1}^k \right) + \in \tag{12}$$

We now prove (10) by induction. Assume that it is true for t = n + 1, n + 2, ⋯, T. But,

$$F_n(s_n) \geq \max_{x^{\pi^k} \in X(s_n)} \left[\sum_{k=1}^{m} \Psi_n^k \left(x_n^{\pi^k} \left(s_{n-1}(b) \right) \right) + Y_{n+1}^{\pi} \left(s_n - \beta \sum_{k=1}^{m} P_k x_{n+1}^k \right) \right], s' = s_{t+1}$$

We now use the induction hypothesis which says $Y_n^{\pi}(s_n) \geq F_n(s_n) - (T-n)\in$ so that

$$\begin{aligned}
F_n(s_n) &\geq \sum_{k=1}^{m} \Psi_n^k \left(x_n^{\pi^k} \left(s_{n-1}(b) \right) \right) + F_{n+1} \left(s_n - \beta \sum_{k=1}^{m} P_k x_{n+1}^k \right) - (T-n-1)\in \\
&\geq \sum_{k=1}^{m} \Psi_n^k \left(x_n^{\pi^k} \left(s_{n-1}(b) \right) \right) + F_{n+1} \left(s_n - \beta \sum_{k=1}^{m} P_k x_{n+1}^k \right) - (T-n-1)\in \\
&\geq \sum_{k=1}^{m} \Psi_n^k \left(x_n^{\pi^k} \left(s_{n-1}(b) \right) \right) + F_{n+1} \left(s_n - \beta \sum_{k=1}^{m} P_k x_{n+1}^k \right) - (T-n)\in + \in \\
&\geq \sum_{k=1}^{m} \Psi_n^k \left(x_n^{\pi^k} \left(s_{n-1}(b) \right) \right) + F_{n+1} \left(s_n - \beta \sum_{k=1}^{m} P_k x_{n+1}^k \right) + \in - (T-n)\in \\
&\geq F_n(s_n) - (T-n)\in
\end{aligned}$$

Hence,

$$\Upsilon_n^*(s_n)+(T-n)\in\geq \Upsilon_n^\pi(s_n)+(T-n)\in$$
$$\geq F_n(s_n)\geq \Upsilon_n^*(s_n)$$

This result shows that solving the optimality equation also gives the optimal value function.

Theorem 1.3: 1) Let B(s) be the set of all bounded real-valued functions F: S → R. The mapping Γ: B(s) → B(s) is a contraction.

2) The operator Γ has a unique fixed point (given by F^*).

3) For any F, $\Gamma^\infty F = F^*$.

4) For any F, if ΓF ≤ F, then $F^* \leq \Gamma^t F$, $t\in\{0,1,\cdots\}\ \forall$.

Note: Γ is called dynamic programming operator (See [5], for more detail).

Theorem 1.4: For any bounded return or scoring functions F_1: S → R and F_2: S → R, and for all t = 0, 1, 2, 3, ···, the inequality below holds

$$\max_{s\in S}\left|(\Gamma^t F_1)(s)-(\Gamma^t F_2)(s)\right|\leq \beta^t \max_{s\in S}\left|F_1(s)-F_2(s)\right|$$

See [8].

The next result shows that as T → ∞, $F^*(s) \to \Gamma^T F(s)$, $\forall\ s\in S$. Thus, the profit per stage must be bounded i.e.

$$\left|\sum_{k=1}^{m}\Psi^k\left(x^{\pi^k}\left(s(b)\right)\right)\right|\leq\mu$$ where μ is a positive constant.

We now state this claim formally as follows:

Theorem 1.5: For any bounded return or scoring function F: S → R, $F_T(s_0) \to F(s_0)$, T → ∞ that is,

$$F(s_0)=\lim_{T\to\infty}\Gamma^T(F)(s_0),\quad \forall\ s_0\in S$$

Proof: Let H be a positive integer, $s_0\in S$ and policy π = {π_0, π_1, ···}, we can decompose the return $F^\pi(s_0)=\lim_{T\to\infty}E\left\{\sum_{t=0}^{T-1}\sum_{k=1}^{m}\beta^t\Psi_t^k\left(x_t^{\pi^k}\left(s_{t-1}(b)\right)\right)\right\}$ into the portion received over the first H stages and over the remaining stages.

$$F^{\pi}(s_0) = \lim_{T\to\infty} E\left\{\sum_{t=0}^{T-1}\sum_{k=1}^{m}\beta^t\Psi_t^k\left(x_t^{\pi^k}\left(s_t(b)\right)\right)\right\}$$
$$= E\left\{\sum_{t=0}^{H-1}\sum_{k=1}^{m}\beta^t\Psi_t^k\left(x_t^{\pi^k}\left(s_t(b)\right)\right)\right\}$$
$$+ \lim_{T\to\infty} E\left\{\sum_{t=H}^{T-1}\sum_{k=1}^{m}\beta^t\Psi_t^k\left(x_t^{\pi^k}\left(s_t(b)\right)\right)\right\}$$

But

$$\left|\lim_{T\to\infty} E\left\{\sum_{t=H}^{T-1}\sum_{k=1}^{m}\beta^t\Psi_t^k\left(x_t^{\pi^k}\left(s_t(b)\right)\right)\right\}\right| \le N\sum_{t=H}^{\infty}\beta^t = \frac{\beta^H N}{1-\beta}$$

Since $\sum_{t=H}^{\infty}\beta^t$ is a geometric progression and $0 < \beta < 1$.

Now

$$F^{\pi}(s_0) \le E\left[\sum_{t=0}^{H-1}\sum_{k=1}^{m}\beta^t\Psi_t^k\left(x_t^{\pi^k}\left(s_t(b)\right)\right) + \frac{\beta^H N}{1-\beta}\right]$$

using this relations, it follows that

$$F^{\pi}(s_0) - \frac{\beta^H N}{1-\beta} - \beta^H \max_{s_t\in S}\left|\Psi(s)\right|$$
$$\le E\left[\beta^H\Psi(s_H) + \sum_{t=0}^{H-1}\sum_{k=1}^{m}\beta^t\Psi_t^k\left(x_t^{\pi^k}\left(s_t(b)\right)\right)\right]$$
$$\le F^{\pi}(s_0) + \frac{\beta^H N}{1-\beta} + \beta^H \max_{s_t\in S}\left|\Psi(s)\right|.$$

By taking the maximum over π, we obtain for all S_0 and H.

$$F^*(s_0) - \frac{\beta^H N}{1-\beta} - \beta^H \max_{s_t\in S}\left|\Psi(s)\right|$$
$$\le \left(\Gamma^H F\right)(s_0) \le F^*(s_0) + \frac{\beta^H N}{1-\beta} + \beta^H \max_{s_t\in S}\left|\Psi(s)\right|,$$

and by taking the unit as $H \to \infty$, we have

$$F^*(s_0) \le \lim_{H\to\infty}\left(\Gamma^H F\right)(s_0) \le F^*(s_0)$$

Hence,

$$F^*(s_0) = \lim_{H \to \infty} \left(\Gamma^H F\right)(s_0), \forall\, s_0 \in S.$$

This result shows that our optimization problem converges to a fixed point F^* in an infinite horizon.

We use value iteration algorithm for finite and infinite stage to solve our problem. The algorithm converges to an optimal policy.

Step 1: Initialization Set $F_0(s_0) = 0 \;\forall\, s_0 \in S.$

Set n = 0 Set $0 < \beta < 1$ Fix a tolerance parameter, $\varepsilon > 0$.

Step 2: For each $s_n \in S$, calculate

$$F_{n+1}(s_n) = \max_{x_n \in X} \sum_{k=1}^{m} \Psi_n^k\left(x_n^k\left(s_n(b)\right)\right) + F_n\left(s_{n-1} - \beta \sum_{k=1}^{m} P_k x_n^k\right) \tag{13}$$

Let x^{n+1} be the decision vector that solve (13).

Step 3: For $\beta = 1$:

If $\|F_{n+1} - F_n\| \le \varepsilon$, set $x^{\varepsilon} = x^{n+1}, F^{\pi^{\varepsilon}} = F_{n+1}$ and stop; else set n = n + 1and return to step 2.

Step 4: For $0 < \beta < 1$;

If $\|F_{n+1} - F_n\| \le \dfrac{1-\beta}{2\beta}$, set $x^{\varepsilon} = x^{n+1}$, $F^{\pi^{\varepsilon}} = F_{n+1}$

and result stop; else set n = n + 1 and return to step 2.

The theorem below guarantees the convergent of the algorithm.

Theorem 1.6: If the algorithm with stopping parameter ε, terminates at iteration n with value function F_{n+1}, then

$$\left\|F_{n+1} - F^*\right\| \le \varepsilon/2 \tag{14}$$

In addition, if x^{ε} is the optimal decision rule and F^{π^k} is the value of this policy, then

$$\left\|F^{\pi^{\varepsilon}} - F^*\right\| \le \varepsilon. \tag{15}$$

This theorem implies that F^{π^k} is the fixed point of the equation $F = \Gamma(x^{\varepsilon})F$. Since x^{ε} is the decision that solves ΓF_{n+1}, it implies that $\Gamma(x^{\varepsilon})F_{n+1} = \Gamma F_{n+1}$. Since $\left\|F^* - F_{n+1}\right\| = \left\|\Gamma F^* - \Gamma F_{n+1} + \Gamma F_{n+1} - F_{n+1}\right\|$ $F_{n+1} = \Gamma F_n$ and Γ is contraction, we have that

$$\left\| F^{\pi^{\epsilon}} - F_{n+1} \right\| \le \frac{\beta}{1-\beta} \left\| F_{n+1} - F_n \right\|$$

and

$$\left\| F_{n+1} - F^{*} \right\| \le \frac{\beta}{1-\beta} \left\| F_{n+1} - F_n \right\| \tag{16}$$

But the value iteration algorithm stops when

$$\left\| F_{n+1} - F_n \right\| \le \varepsilon \frac{(1-\beta)}{2\beta} \tag{17}$$

From (16) and (17), we have

$$\left\| F_{n+1} - F^{*} \right\| \le \frac{\beta}{1-\beta} \times \frac{\varepsilon(1-\beta)}{2\beta} = \varepsilon/2 \tag{18}$$

Similarly, $\left\| F_{n+1} - F^{\pi_{\varepsilon}} \right\| \le \frac{\varepsilon}{2}$

Therefore,

$$\left\| F^{*} - F^{\pi^{\epsilon}} \right\| \le \left\| F_{n+1} - F^{\pi\varepsilon} \right\| + \left\| F^{*} - F_{n+1} \right\| < \frac{\epsilon}{2} + \frac{\epsilon}{2} = \epsilon .$$

COMPUTATIONAL RESULT

A production company in Nigeria proposed to purchase 50 machines that can perform nine different tasks. These machines are subject to breakdown. The **Table 2** gives information of their decisions.

The company further estimated that out of the number of breakdown machines per week, 95% will join the functional ones for the next period.

The aim of the company is to maximize profit over T horizon Let s_t represents the number of machines to be allocated in the next period, so that $s_t = x_t^k(s_{t-1}(b))$. Since s_0 is the number of machines at the beginning of the planning horizon, $\sum_{k=1}^{m} P_k x_t^k$ is the total number of breakdown machines, and $\frac{19}{20}\sum_{k=1}^{m} P_k x_t^k$ is expected to join the functional machines in the next period of operation, we have that $s_t = s_{t-1} - \frac{1}{20}\sum_{k=1}^{m} P_k x_t^k, t = T, T-1, \cdots, 1$ which is our transformation equation and is a random variable.

We can now express our one-period expected return function as follows:

$$\Upsilon_t^{\pi}(s_t) = E\left[\max_{x_t^k \in X} \sum_{t=1}^{T} \sum_{k=1}^{m} \beta^t \Psi_t^k \left(x_t^k \left(s_{t-1}(b) \right) \right) \right] \tag{19}$$

subject to

$$\sum_{k=1}^{m} x_t^k\left(s_{t-1}(b)\right) = s_{t-1}, t = T, T-1, \cdots, 1$$

(The feasible region for stage t).

Since the company cannot allocate negative resources to any one task, we write $x_t^k\left(s_{t-1}(b_t)\right) \geq 0$, t = T, T – 1, ···, 1; k = 1, ···, m.

Of course, (19) is the same as

$$F_T\left(s_{t-1}\right) = \max_{x_t^k \in X}\left\{\sum_{k=1}^{m} \Psi_t^k x_t^{\pi^k}\left(s_{t-1}\left(b_t\right)\right) + EF_{T-1}\left(s_t\right)\right\},$$

$t = T, T-1, \cdots, 1,$

$$= \max_{x_t^k \in X}\left[\sum_{k=1}^{m} \Psi_t^k\left(x_t^{\pi^k}\left(s_{t-1}\left(b_t\right)\right)\right) + EF_{T-1}\left(s_{t-1} - \frac{1}{20}\sum_{k=1}^{m} P_k x_t^k\right)\right],$$

t = T, T – 1, ···, 1, and $x_t^k \geq 0$, t = T, T – 1, ···, 1; k = 1, ···,m.

Note: Our problem has 9 tasks and 48 periods.

Hence, m = 9 and T = 48. Set $\varepsilon = \frac{1}{9}$.

Therefore,

$$\begin{aligned} &F_T\left(s_{t-1}\right) \\ &= \max_{x^k \in X}\Big\{180000x_t^1 + 150000x_t^2 + 192000x_t^3 + 240000x_t^4 \\ &\quad + 228000x_t^5 + 120000x_t^6 + 168000x_t^7 + 222000x_t^8 \\ &\quad + 300000x_t^9 \\ &\quad + F_{T-1}\Big(s_{t-1} - \frac{1}{20}\Big[\frac{2}{11}x_t^1 + \frac{1}{10}x_t^2 + \frac{2}{13}x_t^3 + \frac{5}{21}x_t^4 \\ &\qquad + \frac{3}{13}x_t^5 + \frac{1}{12}x_t^6 + \frac{1}{9}x_t^7 \\ &\qquad + \frac{4}{15}x_t^8 + \frac{5}{18}x_t^9\Big]\Big)\Big\} \end{aligned}$$

$$T = 1, 2, \cdots, 48; t = 48, 47, \cdots, 1. \tag{20}$$

subject to:

$$\sum_{k=1}^{9} x_t^k = S_{t-1}, t = 48, 47, \cdots, 1,$$

$x_t^k \geq 0$, t = 48, 47, ···, 1; k = 1, ···, 9 which is a parametric linear programming problem with 9 variables.

A program using MatLab was used for (20). At the end, the following results were obtained.

The profit over 48 weeks is given by

Table 2: The Expected Initial Profit and Probability of Breakdown Machines.

Machines	Task x^1	Task x^2	Task x^3	Task x^4	Task x^5	Task x^6	Task x^7	Task x^8	Task x^9
Initial Expected Profit (in Naira)	180,000	150,000	192,000	240,000	228,000	120,000	168,000	222,000	300,000
Probability of Breakdown (P_k)	2/11	1/10	2/13	5/21	3/19	1/12	1/9	4/15	5/18

$$\begin{aligned} F_{48}(s_0) &= ₦91422000 s_o \\ &= ₦91422000 \times 50 \\ &= ₦4,571,100,000. \end{aligned}$$

By theorem 1.5, as T approach infinity, $F_T(s_0)$ approaches $F(s_0)$,

$$\begin{aligned} F(s_0) &= \lim_{T\to\infty} \left(\Gamma^T F\right)(s_0) \\ &= \lim_{T\to\infty} F_T(s_0) = ₦409820000 s_0 \\ &= ₦20,491,000,000. \end{aligned}$$

and the optimal policy $x_t^9(s_0)$, (t = 1, 2, ···), approaches the limit zero.

Discussion: **Figure 1** shows the expected profit that will accrued to the company over a period of 48 weeks. We found that at 48 weeks, the maximum profit that accrued to the company to be ₦4,571,100,000. **Figure 2** shows the expected profit that will accrued to the company over an infinite weeks. It was found that at infinity, the maximum profit that will accrued to the company to be ₦20,491,000,000.

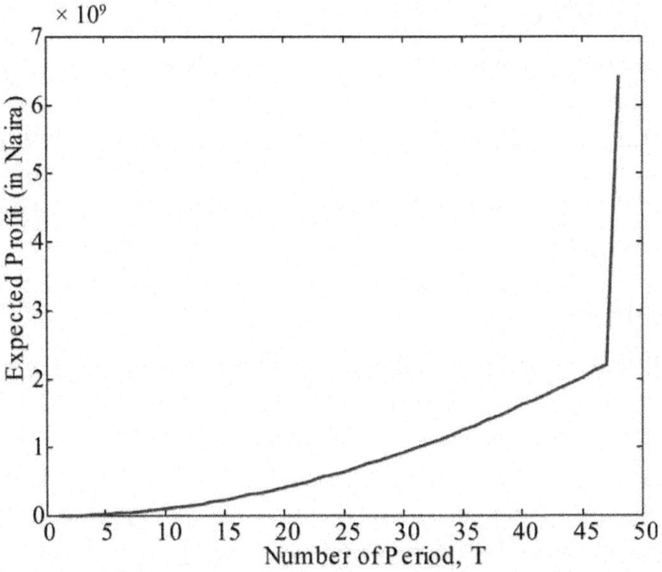

Figure 1: The Expected Profit that will accrued to the Company over a Period of 48 weeks.

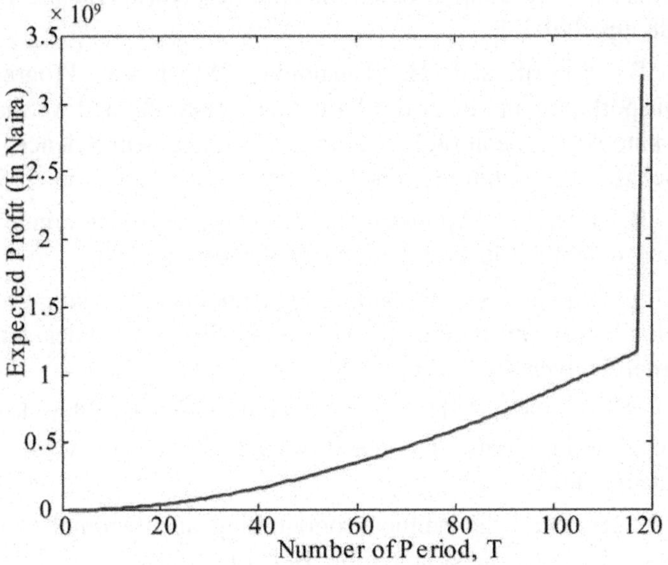

Figure 2: The Expected Profit that will accrued to the Company over an Infinite Period.

CONCLUSION

Many production companies have for long been allocating resources to different tasks without putting into consideration certain factors that may hinder the realization of their objectives. This paper dealt with allocation of machines to tasks in order to maximize profit over finite and infinite horizon. Careful analysis of the situation reveals that adequate planning, prompt and effective maintenances can enhances the profitability of the company.

REFERENCES

1. J. M. Mulvey and H. Vladimirou, "Stochastic Network Programming for Financial Planning Problems," Management Science, Vol. 38, No. 11, 1992, pp. 1642-1664.doi:10.1287/mnsc.38.11.1642
2. B. Van Roy, D. P. Bertsekas, Y. Lee and J. N. Tsitsiklis, "A Neuro-Dynamic Programming Approach to Retailer Inventory Management," Proceedings of the 36th IEEE Conference on Decision and Control, San Diego, 10-12 December 1997, pp. 4052-4057.doi:10.1109/CDC.1997.652501
3. W. B. Powell, "A Comparative Review of Alternative Algorithms for the Dynamic Vehicle Allocation Problem," In: B. Golden and A. Assad, Eds., Vehicle Routing: Methods and Studies, North Holland, Amsterdam, 1988, pp. 249-292.
4. W. B. Powell and H. Topaloglu, "Stochastic Programming in Transportation and Logistics," In: A. Ruszczynski and A. Shapiro, Eds., Handbook in Operation Research and Management Science, Volume on Stochastic Programming, Elsevier, Amsterdam, 2003, pp. 555-635.
5. W. B. Powell, "Approximate Dynamic Programming for Asset Management," Princeton University, Princeton, 2004.
6. H. Topaloglu and S. Kunnumkal, "Approximate Dynamic Programming Methods for an Inventory Allocation Problem under Uncertainty," Cornell University, Ithaca, 2006.
7. C. I. Nkeki, "On A Dynamic Programming Algorithm for Resource Allocation Problems," Unpublished M.Sc. Thesis, University of Ibadan, Ibadan, 2006.
8. D. P. Bertsekas, "Dynamic Programming and Optimal Control," 2nd Edition, Vol. 2, Athena Scientific, Belmont, 2001.

Chapter 12

A DIFFERENT APPROACH TO CONE-CONVEX OPTIMIZATION

Surjeet Kaur Suneja[1], Sunila Sharma[1], Meetu B. Grover[1], Malti Kapoor[2]

[1]Department of Mathematics, Miranda House College, University of Delhi, Delhi, India

[2]Department of Mathematics, Moti Lal Nehru College (M), University of Delhi, Delhi, India

ABSTRACT

In classical convex optimization theory, the Karush-Kuhn-Tucker (KKT) optimality conditions are necessary and sufficient for optimality if the objective as well as the constraint functions involved is convex. Recently, Lassere [1] considered a scalar programming problem and showed that if the convexity of the constraint functions is replaced by the convexity of the feasible set, this crucial feature of convex programming can still be preserved. In this paper, we generalize his results by making them applicable to vector optimization problems (VOP) over cones. We consider the minimization of a cone-convex function over a convex feasible set described by cone constraints that are not necessarily cone-convex. We show that if a Slater-type cone constraint qualification holds, then every weak minimizer of (VOP) is a KKT point and conversely every KKT point is a weak minimizer. Further a Mond-Weir type dual is formulated in the modified situation and various duality results are established.

INTRODUCTION

Convex programming deals with the minimization of a convex objective function over a convex set usually described by convex constraint functions. In the past various attempts have been made to weaken the convexity hypothesis [2-4] by replacing convex objective as well as constraint functions with more general ones and thus exploring the extent of optimality conditions applicability.

As a breakthrough to this, Lassere [1] showed that as far as KKT optimality conditions are concerned, the convexity (or any of its generalization) of the constraint functions can be replaced by the convexity of the feasible set described by the constraints. More precisely, Lassere considered the following convex optimization problem (CP):

(CP) minimize $f(x)$ subject to $g_j(x) \le 0,\ j = 1,\cdots,m$

where $f: R^n \to R$ is a differentiable convex function and the feasible set $F_0 = \{x \in R^n : g_j(x) \le 0, j = 1,\cdots,m\}$

is a convex set while the $g_j's$: $R^n \to R$ are differentiable but not necessarily convex functions. To prove the necessity and sufficiency of KKT conditions in this framework Lassere considered the following non-degeneracy condition (ND$_1$): For all $j = 1,\cdots,m$,

$\nabla g_j(x) \ne 0$, whenever $x \in F_0$ and $g_j(x) = 0$ (ND$_1$)

He showed that if the Slater constraint qualification1 and the above non-degeneracy condition (ND$_1$) hold, then a feasible point x* of (CP) is a global minimizer if and only if it is a KKT point, that is,

$$\nabla f(x^*) + \sum_{j=1}^{m} \lambda_j \nabla g_j(x^*) = 0$$
,

And $\lambda_j g_j(x^*) = 0$, $j = 1,\cdots,m$ (KKT$_1$) for some non-negative vector $\lambda = (\lambda_1,\cdots,\lambda_m) \in R^m$.

This work of Lassere [1] has been carried forward to the non-smooth case by Dutta and Lalitha [5]. They considered the same problem (CP) with the only difference being that the function f is a non-differentiable convex function and the convex set F_0 is described by local Lipschitz constraint functions g_j which are not necessarily differentiable or convex. In terms of Dutta and Laltha [5] a point $x^* \in F_0$ is said to be a KKT point for the problem (CP) if there exist scalars $\lambda_j \ge 0, j = 1,\cdots,m$, such that $0 \in \partial f(x^*) + \sum_{j=1}^{m} \lambda_j \partial^0 g_j(x^*)$, and $\lambda_j g_j(x^*) = 0$, $j = 1,\cdots,m$ (KKT$_2$)

where

$\partial f(x^*) = \{\xi \in R^n : f(y) - f(x^*) \ge \langle \xi, y - x^* \rangle, \forall y \in R^n\}$ denotes the sub-differential of f at x* and $\partial^0 g_j(x^*) = \{\xi \in R^n : g_i^0(x,d) \ge \langle \xi, d \rangle, \forall d \in R^n\}$ denotes the Clarke sub-differential of the function g_j at x*.

Further, Dutta and Lalitha [5] introduced the following non-smooth version (ND_2) of Lassere's non-degeneracy condition:

For all $j = 1, \cdots, m$

$$0 \notin \partial^0 g_j(x), \text{ whenever } x \in F_0 \text{ and } g_j(x) = 0 \quad (ND_2)$$

In this modified setting Dutta and Lalitha [5] concluded that if each g_j is assumed to be regular in the sense of Clarke [6] and if the Slater constraint qualification and the non-degeneracy condition (ND_2) hold, then a feasible point x^* is a global minimizer of f over F_0 if and only if it is a KKT point.

The overall aim of this paper is to extend Lassere's [1] results to a vector optimization problem over cones.

PRELIMINARIES AND PROBLEM FORMULATION

We consider the following vector optimization problem (VOP) over cones:

(VOP) K – minimize $f(x)$ subject to $g(x) \in Q$

where $f : R^n \to R^p$ and $g : R^n \to R^m$ are differentiable functions, K and Q are closed convex cones with non-empty interiors in R^p and R^m respectively.

Let $F = \{x \in R^n : -g(x) \in Q\}$ be the set of feasible solutions of (VOP).

The positive dual cone K^* and the strict positive dual cone K^{s*} of K are respectively defined as $K^* = \{z \in R^p : x'z \geq 0 \text{ for all } x \in K\}$ and $K^{s*} = \{z \in R^p : x'z > 0 \text{ for all } x \in K \setminus \{0\}\}$.

We begin by defining the notion of a KKT point in terms of (VOP).

Definition 2.1: A point $x^* \in F$ is said to be a KKT-point if there exist $\lambda \in K^* \setminus \{0\}$ and $\mu \in Q^*$ such that

$\lambda' \nabla f(x^*) + \mu' \nabla g(x^*) = 0$ and $\mu' g(x^*) = 0$.

For the problem (VOP), the solutions are defined in the following sense:

Definition 2.2 [7]: A point $x^* \in F$ is called 1) a weak minimum of (VOP) if for all $x \in F$

$f(x^*) - f(x) \notin \operatorname{int} K$;

2) a Pareto-minimum of (VOP) if for all $x \in F$

$f(x^*) - f(x) \notin K \setminus \{0\}$;

3) a Strong minimum of (VOP) if for all $x \in F$

$f(x)-f(x^*)\in K$.

Let F^w denote the set of weak minimum solutions of (VOP).

The forthcoming optimality and duality results are based on suitable generalized convexity assumptions over cones, thus we recall some known definitions in the literature.

Definition 2.3 [8,9]: A function $f:R^n\rightarrow R^p$ is said to be

1) K-convex at a point $x^*\in R^n$ if for every $x\in R^n$

$f(x)-f(x^*)-\nabla f(x^*)(x-x^*)\in K$.

2) K-pseudoconvex at $x^*\in R^n$ if for every $x\in R^n$

$-\nabla f(x^*)(x-x^*)\notin \operatorname{int} K \Rightarrow f(x^*)-f(x)\notin \operatorname{int} K$.

3) strongly K-pseudoconvex at $x^*\in R^n$ if for every $x\in R^n$

$-\nabla f(x^*)(x-x^*)\notin \operatorname{int} K \Rightarrow f(x)-f(x^*)\in K$.

4) strictly K-pseudoconvex at $x^*\in R^n$ if for every $x\in R^n$

$-\nabla f(x^*)(x-x^*)\notin \operatorname{int} K \Rightarrow f(x^*)-f(x)\notin K$.

If f is K-convex (K-pseudoconvex, strongly K-pseudoconvex, strictly K-pseudoconvex) at every $x^*\in R^n$ then f is said to be K-convex (K-pseudoconvex, strongly K-pseudoconvex, strictly K-pseudoconvex) on R^n.

On the lines of Jahn [10] we define Slater-type cone constraint qualification as follows:

Definition 2.4: The problem (VOP) is said to satisfy Slater-type cone constraint qualification at $x^*\in F$ if there exists $\hat{x}\in R^n$ such that

$g(x^*)+\nabla g(x^*)(\hat{x}-x^*)\in -\operatorname{int} Q$.

Note that if g is Q-convex at x^* and the problem (VOP) satisfies Slater constraint qualification, that is, there exists $\hat{x}\in R^n$ such that $-g(\hat{x})\in \operatorname{int} Q$, then (VOP) satisfies Slater-type cone constraint qualification at x^*.

Also, it is worth noticing that following the steps of Lassere [1] and Dutta and Lalitha [5] we can define the analogous non-degeneracy condition (ND_3) for (VOP) as follows:

For all $\mu\in Q^*\setminus\{0\}$, $\mu'\nabla g(x)\neq 0$, whenever $x\in F$ and $\mu' g(x)=0$.

But if we assume that Slater-type cone constraint qualification holds at a point $x^* \in F$, then there exists $\hat{x} \in R^n$ such that $g(x^*)+\nabla g(x^*)(\hat{x}-x^*) \in -\text{int}\,Q$ which means that for all $\mu \in Q^* \setminus \{0\}$ for which

$\mu' g(x^*)=0$, we have $\mu' \nabla g(x^*)(\hat{x}-x^*)<0$ which itself implies that $\mu' \nabla g(x^*) \neq 0$ and hence the nondegeneracy condition holds.

Thus in the paper, we shall extend Lassere's [1] results to the vector optimization problem (VOP) over cones but, unlike Lassere, to prove our results we need to assume only Slater-type cone constraint qualification at a point.

OPTIMALITY CONDITIONS

In this section we prove several classical optimality results by taking generalized convexity assumptions over cones on the objective function and assuming the feasible set to be convex and with no convexity type restriction on the constraint function. It is clear that if the constraint function g in (VOP) is Q-convex then the feasible set F is convex, so we begin by exemplifying the fact that F can be convex without g being Q-convex.

Example 3.1: Consider $g: R \to R^2$ defined as

$g(x)=(x,\ 6-x^2)$ and $Q=\{(x,y): y \geq x, x \leq 0\}$.

Here g is not Q-convex, because if we take $x=5/2$ and $x^*=3$ then

$$g(x)-g(x^*)-\nabla g(x^*)(x-x^*)=\left(0,-\frac{1}{4}\right) \notin Q$$

But the feasible set $F=\{x \in R: x \geq 2\}$ is convex. We have the following lemma.

Lemma 3.1: If the feasible set F of (VOP) is convex then

$$(\mu' \nabla g(x^*))(x-x^*) \leq 0, \text{for all } x \in F \tag{1}$$

where $x^* \in F, \mu \in Q^* \setminus \{0\}$ satisfy $\mu' g(x^*)=0$.

Proof: Let F be convex and suppose

$x^* \in F, \mu \in Q^* \setminus \{0\}$ satisfy $\mu' g(x^*)=0$.

Assume that

$$(\mu' \nabla g(x^*))(x-x^*)>0 \text{ for some } x \in F \tag{2}$$

Now, for $0<\alpha<1$, we have

$$g\left(x^*+\alpha\left(x-x^*\right)\right)$$
$$=g\left(x^*\right)+\nabla g\left(x^*\right)\alpha\left(x-x^*\right)+o(\alpha)$$

where

$$\lim_{\alpha\to 0^+} o(\alpha)=0.$$

This implies that.

$$\mu' g\left(x^*+\alpha\left(x-x^*\right)\right)$$
$$=\mu' g\left(x^*\right)+\alpha\left(\mu'\nabla g\left(x^*\right)\right)\left(x-x^*\right)+\mu' o(\alpha)$$

Using (2) together with $\mu' g\left(x^*\right)=0$ for a sufficiently small, $\alpha>0$, we get

$$\mu' g\left(x^*+\alpha\left(x-x^*\right)\right)>0. \quad (3)$$

Since F is convex, therefore $x^*+\alpha\left(x-x^*\right)\in F$, that is, $-g\left(x^*+\alpha\left(x-x^*\right)\right)\in Q$ so that $\mu' g\left(x^*+\alpha\left(x-x^*\right)\right)\leq 0$.

This contradicts (3). Hence the result.

The above lemma plays a pivotal role throughout the rest of the paper, thus we illustrate it by means of an example.

Example 3.2: Consider $g:R\to R^2$ and Q as defined in Example 3.1. Then we have already seen that g is not Q-convex whereas the feasible set F is convex.

Now, if we take $x^*=2\in F$, then $\mu' g\left(x^*\right)=0$ if and only if $\mu=(-\alpha,\alpha),\alpha\geq 0$, and for this choice of m,

$$\left(\mu'\nabla g\left(x^*\right)\right)\left(x-x^*\right)=-5\alpha(x-2)\leq 0 \text{ for all } x\in F$$

Also, for any other $x^*>2$, there does not exist any $\mu\in Q^*\setminus\{0\}$ for which $\mu' g\left(x^*\right)=0$.

Hence the lemma holds.

The following theorem serves the main purpose of the paper.

Theorem 3.1: Consider a feasible solution x* of the vector optimization problem (VOP) and assume that Slater-type cone constraint qualification holds at x*. If f is K-convex at x* and the feasible set F is convex then x* is a weak minimum of (VOP) if and only if it is a KKTpoint.

Proof: Let $x^* \in F$ be a weak minimum of (VOP). By Lemma 1 [11], there exist $\lambda \in K^*$ and $\mu \in Q^*$ not both zero such that

$$\left(\lambda^t \nabla f\left(x^*\right) + \mu^t \nabla g\left(x^*\right)\right)\left(x - x^*\right) \geq 0, \forall x \in R^n \tag{4}$$

and

$$\mu^t g\left(x^*\right) = 0. \tag{5}$$

If possible, let $\lambda = 0$, then $\mu \neq 0$ so that from (4), we get

$$\left(\mu^t \nabla g\left(x^*\right)\right)\left(x - x^*\right) \geq 0, \forall x \in R^n. \tag{6}$$

Since Slater-type cone constraint qualification holds at x*, there exists $\hat{x} \in R^n$ such that

$g\left(x^*\right) + \nabla g\left(x^*\right)\left(\hat{x} - x^*\right) \in -\operatorname{int} Q$ which gives that

$$\mu^t \left\{g\left(x^*\right) + \nabla g\left(x^*\right)\left(\hat{x} - x^*\right)\right\} < 0.$$

This together with (5) implies

$\left(\mu^t \nabla g\left(x^*\right)\right)\left(\hat{x} - x^*\right) < 0$ which contradicts (6). Therefore $\lambda \neq 0$.

Since the inequality (4) holds for every $x \in R^n$, we conclude that

$$\lambda^t \nabla f\left(x^*\right) + \mu^t \nabla g\left(x^*\right) = 0 \tag{7}$$

and

$$\mu^t g\left(x^*\right) = 0. \tag{8}$$

Hence x* is a KKT-point.

Conversely, let $x^* \in F$ be a KKT-point, that is, there exist $\lambda \in K^* \setminus \{0\}$ and $\mu \in Q^*$ such that (7) and (8) hold.

Suppose x* is not a weak minimum of (VOP), so there exists $\hat{x} \in F$ such that

$$f\left(x^*\right) - f\left(\hat{x}\right) \in \operatorname{int} K. \tag{9}$$

Since f is K-convex at x*,

$$f\left(\hat{x}\right) - f\left(x^*\right) - \nabla f\left(x^*\right)\left(\hat{x} - x^*\right) \in K. \tag{10}$$

By (9) and (10),

$-\nabla f(x^*)(\hat{x}-x^*) \in \operatorname{int} K$ which implies $(\lambda' \nabla f(x^*))(\hat{x}-x^*) < 0$.

This, by (7), gives

$$(\mu' \nabla g(x^*))(\hat{x}-x^*) > 0.$$

But this contradicts Lemma 3.1 as $\mu' g(x^*) = 0$.

Hence x^* is a weak minimum for (VOP).

Theorem 3.2: Let f be K-pseudoconvex at $x^* \in F$ and suppose that F is convex. Further assume that Slater-type cone constraint qualification holds at x^*. Then x^* is a weak minimum of (VOP) if and only if it is a KKT-point.

Proof: Proof follows on similar lines as Theorem 3.1.

Now we obtain sufficient optimality conditions for (VOP).

Theorem 3.3: Let f be K-convex at $x^* \in F$ and the feasible set F be convex and suppose that there exist $\lambda \in K^{s*}$ and $\mu \in Q^*$ such that (7) and (8) hold. Then x^* is a Pareto minimum of (VOP).

Proof: Let if possible, x^* be not a Pareto minimum of (VOP). Then there exists $\hat{x} \in F$ such that

$$f(x^*) - f(\hat{x}) \in K \setminus \{0\}. \tag{11}$$

Since f is K-convex at x^*, we have

$$f(\hat{x}) - f(x^*) - \nabla f(x^*)(\hat{x}-x^*) \in.$$

Using (11), we get

$$-\nabla f(x^*)(\hat{x}-x^*) \in K \setminus \{0\} K.$$

Since $\lambda \subset K^{s*}$, we have

$$(\lambda' \nabla f(x^*))(\hat{x}-x^*) < 0.$$

Now proceeding as in the converse part of Theorem 3.1, we get a contradiction to Lemma 3.1. Hence x^* is a Pareto minimum of (VOP).

We now give an example to illustrate Theorem 3.3.

Example 3.3: Consider the problem

(VOP) K-Minimize $f(x)$

Subject to $-g(x) \in Q$

where $g: R \to R^2$ and Q are as defined in Example 3.1 and $f: R \to R^2$ and K are given by

$$f(x) = (x^2 + 2x - 3, -x - 3), K = \{(x, y) : -x \le y \le x\}.$$

Then, as shown in Example 3.1, g is not Q-convex. while the feasible set $F = \{x \in R : x \ge 2\}$ of (VOP) is convex. Also f is K-convex at $x^* = 2$.

It can be seen that for

$$\lambda = \left(1, \frac{3}{4}\right) \in K^{s*} \text{ and } \mu = \left(\frac{-21}{20}, \frac{21}{20}\right) \in Q^*,$$

$\lambda' \nabla f(x^*) + \mu' \nabla g(x^*) = 0,$ and $\mu' g(x^*) = 0$.

Thus by Theorem 3.3, $x^* = 2$ is a Pareto minimum of (VOP).

Remark 3.1: Example 3.3 describes a vector optimization problem in which a Pareto minimum is obtained by applying Theorem 3.3 whereas it is impossible to do so using Lassere's [1] results.

Theorem 3.4: Let f be strictly K-pseudoconvex at $x^* \in F$ and the feasible set F be convex and suppose that there exist $\lambda \in K^* \setminus \{0\}$ and $\mu \in Q^*$ such that (7) and (8) hold. Then x^* is a Pareto minimum of (VOP).

Proof: Let if possible, x^* be not a Pareto minimum of (VOP).

Then there exists $\hat{x} \in F$ such that $f(x^*) - f(\hat{x}) \in K \setminus \{0\}$.

Since f is strictly K-pseudoconvex at x^*, we get

$$-\nabla f(x^*)(\hat{x} - x^*) \in \operatorname{int} K.$$

As $\lambda \in K^* \setminus \{0\}$, we have

$$(\lambda' \nabla f(x^*))(\hat{x} - x^*) < 0.$$

Now proceeding as in the converse part of Theorem 3.1, we get a contradiction to Lemma 3.1. Hence x^* is a Pareto minimum of (VOP).

Theorem 3.5: Let f be strongly K-pseudoconvex at $x^* \in F$ and the feasible set F be convex and suppose that there exist $\lambda \in K^* \setminus \{0\}$ and $\mu \in Q^*$ such that (7) and (8) hold. Then x^* is a strong minimum of (VOP).

Proof: Let if possible, x^* be not a strong minimum of (VOP).

Then there exists $\hat{x} \in F$ such that $f(\hat{x}) - f(x^*) \notin K$.

Since f is strongly K-pseudoconvex at x^*, we get

$$-\nabla f(x^*)(\hat{x} - x^*) \in \operatorname{int} K.$$

As $\lambda \in K^* \setminus \{0\}$, we have

$$(\lambda' \nabla f(x^*))(\hat{x} - x^*) < 0.$$

Again proceeding as in the converse part of Theorem 3.1, we get a contradiction. Hence x^* is a strong minimum of (VOP).

DUALITY

With the primal problem (VOP), we associate the following Mond-Weir type dual program (MDP):

(MDP) K-maximize $f(y)$

subject to

$$(\lambda' \nabla f(y) + \mu' \nabla g(y))(x - y) \geq 0, \forall x \in F \tag{12}$$

$$\mu' g(y) = 0, \tag{13}$$

$$y \in F, \lambda \in K^* \setminus \{0\}, \mu \in Q^*.$$

Let F^D denote the set of feasible solutions of (MDP).

Definition 4.1: A point $(\bar{y}, \bar{\lambda}, \bar{\mu}) \in F^D$ is said to be a weak maximum of (MDP) if

$$f(y) - f(\bar{y}) \notin \operatorname{int} K, \text{for all } (y, \lambda, \mu) \in F^D.$$

Let F_W^D denote the set of weak maximum solutions of (MDP).

Theorem 4.1: (Weak Duality) Let $x \in F$ and $(y, \lambda, \mu) \in F^D$. Assume that f is K-pseudoconvex at y and the feasible set F is convex, then

$$f(y) - f(x) \notin \operatorname{int} K.$$

Proof: Let $x \in F$ and $(y, \lambda, \mu) \in F^D$. Suppose to the contrary that

$$f(y) - f(x) \in \operatorname{int} K. \tag{14}$$

Since f is K-pseudoconvex at y, (14) implies

$$-\nabla f(y)(x-y) \in \operatorname{int} K.$$

As $\lambda \in K^* \setminus \{0\}$, we get

$$(\lambda' \nabla f(y))(x-y) < 0. \tag{15}$$

Since $\mu' g(y) = 0$, therefore by Lemma 3.1,

$$(\mu' \nabla g(y))(x-y) \leq 0. \tag{16}$$

Adding (15) and (16), we have $(\lambda' \nabla f(y) + \mu' \nabla g(y))(x-y) < 0$ which contradicts (12). Hence, $f(y) - f(x) \notin \operatorname{int} K$.

Theorem 4.2: (Strong Duality) Let $x^* \in F^W$. Assume that Slater-type cone constraint qualification holds at x*. If f is K-pseudoconvex at x* and the feasible set F is convex, then there exist $\lambda^* \in K^* \setminus \{0\}$ and $\mu^* \in Q^*$ such that $(x^*, \lambda^*, \mu^*) \in F^D$. Further, if the conditions of Weak Duality Theorem 4.1 hold for all $x \in F$ and $(y, \lambda, \mu) \in F^D$, then $(x^*, \lambda^*, \mu^*) \in F_W^D$.

Proof: Since all the conditions of Theorem 3.2 hold, therefore there exist $\lambda^* \in K^* \setminus \{0\}$ and $\mu^* \in Q^*$ such that

$$\lambda^{*\prime} \nabla f(x^*) + \mu^{*\prime} \nabla g(x^*) = 0 \text{ and } \mu^{*\prime} g(x^*) = 0.$$

Thus $(x^*, \lambda^*, \mu^*) \in F^D$. Further if $(x^*, \lambda^*, \mu^*) \notin F_W^D$, then there exists $(y, \lambda, \mu) \in F^D$ such that $f(y) - f(x^*) \in \operatorname{int} K$, which contradicts Theorem 4.1.

Hence, $(x^*, \lambda^*, \mu^*) \in F_W^D$.

Theorem 4.3: (Converse Duality) Let

$$(y^*, \lambda^*, \mu^*) \in F_W^D.$$

Assume that f is K-pseudoconvex at y^* and the feasible set F is convex. Then $y^* \in F^W$.

Proof: Suppose $y^* \notin F^W$. Then there exists $\hat{x} \in F$ such that

$$f(y^*) - f(\hat{x}) \in \operatorname{int} K.$$

Since f is K-pseudoconvex at y^*, we get

$-\nabla f(y^*)(\hat{x}-y^*) \in \text{int}\, K$ so that,

$$\left(\lambda^{*t}\nabla f(y^*)\right)(\hat{x}-y^*) < 0. \tag{17}$$

Also, $\mu^{*t} g(y^*) = 0$, so that by Lemma 3.1,

$$\left(\mu^{*t}\nabla g(y^*)\right)(\hat{x}-y^*) \le 0 \tag{18}$$

Adding (17) and (18), we have

$\left(\lambda^{*t}\nabla f(y^*) + \mu^{*t}\nabla g(y^*)\right)(\hat{x}-y^*) < 0$ which contradicts (12). Hence, $y^* \in F^W$.

CONCLUSION

This paper gives a new direction to the search for solution of a vector optimization problem over cones. We have shown that, with Slater-type cone constraint quailfication, convexity of the feasible set can replace the cone-convexity (or any of its generalization) of the constraint functions, and then we just need to assume the cone-convexity (or a suitable generalization) of the objective function to prove the necessity and sufficiency of the KKT optimality conditions. Moreover, a Mond-Weir type dual has been formulated in the modified situation and various duality results have been established.

ACKNOWLEDGEMENTS

The first author is grateful to the University Grants Commission (UGC), India for offering financial support.

REFERENCES

1. J. B. Lasserre, "On Representations of the Feasible Set in Convex Optimization," Optimization Letters, Vol. 4, No. 1, 2010, pp. 1-5. http://dx.doi.org/10.1007/s11590-009-0153-6
2. C. R. Bector, S. Chandra and M. K. Bector, "Sufficient Optimality Conditions and Duality for a Quasiconvex Programming Problem," Journal of Optimization Theory and Applications, Vol. 59, No. 2, 1988, pp. 209-221.
3. O. L. Mangasian, "Non-Linear Programming," McGraw Hill, New York, 1969.
4. S. Nobakhtian, "Sufficiency in Nonsmooth Multiobjective Programming Involving Generalized (F, r)-Convexity," Journal of Optimization Theory

and Applications, Vol. 130, No. 2, 2006, pp. 359-365. http://dx.doi.org/10.1007/s10957-006-9105-9

5. J. Dutta and C. S. Lalitha, "Optimality Conditions in Convex Optimization Revisited," Optimization Letters, Vol. 7, No. 2, 2013, pp. 221-229.http://dx.doi.org/10.1007/s11590-011-0410-3
6. F. H. Clarke, "Optimization and Non-smooth Analysis," Wiley, New York, 1983.
7. L. Coladas, Z. Li and S. Wang, "Optimality Conditions for Multiobjective and Nonsmooth Minimization in Abstract Spaces," Bulletin of Australian Mathematical Society, Vol. 50, No. 2, 1994, pp. 205-218. http://dx.doi.org/10.1017/S0004972700013678
8. S. Aggarwal, "Optimality and Duality in Mathematical Programming Involving Generalized Convex Functions," Ph.D. Thesis, University of Delhi, Delhi, 1998.
9. T. Weir, B. Mond and B. D. Craven, "Weak Minimization and Duality," Numerical Functional Analysis and Optimization, Vol. 9, No. 1-2 ,1987, pp. 181-192.
10. J. Jahn, "Vector Optimization," Springer Verlag, New York, 2003.
11. S. K. Suneja, S. Aggarwal and S. Davar, "Multiobjective Symmetric Duality Involving Cones," European Journal of Operational Research, Vol. 141, No. 3, 2002, pp. 471- 479.http://dx.doi.org/10.1016/S0377-2217(01)00258-2

Chapter 13

PREDICTION AND OPTIMIZATION OF SYSTEM QUALITY AND RISKS ON THE BASE OF MODELLING PROCESSES

Andrey Kostogryzov[1], Leonid Grigoriev[2], George Nistratov[3], Andrey Nistratov[4], Vladimir Krylov[5]

[1]Research Institute of Applied Mathematics and Certification, Moscow, Russia

[2]Gubkin Russian State University of Oil and Gas, Moscow, Russia

[3]Research Institute of Applied Mathematics and Certification, Moscow, Russia

[4]Institute of Informatic Problems of the Russian Academy of Sciences, Moscow, Russia

[5]Scientific & Technical Center "NAOPRO", Moscow, Russia

ABSTRACT

The paper is concerned with the development and application of the original probability models and supporting them software tools to predict and optimize quality and risks for complex systems. The examples demonstrate possibilities to use modeling results from different application spheres and to go in making decision "from a pragmatical filtration of information to generation of the proved ideas and effective decisions".

INTRODUCTION

Today system processes for different conditions and threats are the main objects for improvement of system operation. The goal of this work is to propose models, methods, and software tools well-tested in practice, to predict and optimize quality and risks as applied to newly developed and currently operated manufacture, power generation, transport, engineering, information, control and measurement, food storage, quality assurance and security systems. Presented work covers logically closed contour: "system requirements—supporting mathematical models to estimate processes and system operation—

ways to optimize quality and risks". Thereby the answers on system engineering questions—"Is expected quality achievable?", "Can be the system requirements met?", "How much safe are those or others scenarios?", "What about the real risks, profits and possible damages?", "What choice in system architecture is rational?", "What analyzed variants and decision are more effective and why?", "What rational measures should lead to estimated effect without waste expenses, when, by which controllable and uncontrollable conditions and costs?" etc.— can be substantiated quantitatively. The answers may be received before critical events (not only after these events). As demonstration 10 practical examples are investigated and explained, the detailed 'hardware' of the work (including dozens of models), other hundreds examples and routine comments are gathered at [1-12] and on site www.mathmodels.net.

Why you should trust to the results of prediction by the offered models? In other words how models adequacy is substantiated? Though any answer to these questions won't be irrefragable for a certain system we shall try to formulate our arguments (experience readers understand that any model needs in similar arguments).

Argument 1: The fact is that while shaping models all mathematical results are initially drawn in the integral form. As input data are somehow connected with time after choosing probability distribution functions (PDF) characterizing these data there were selected the gamma distribution and the Erlang's distribution. Mathematicians know that these distributions approximate sums of positively distributed random variables well. Every temporary data are as a matter of fact such a sum of compound time expenses. Studies of regularities (for example, [13-18]) have shown that extremes are achieved on bounds of these distributions, i.e. of exponential and deterministic (discrete) distributions. Thus, real values will be somewhere between lower and upper estimations. The results reflect pessimistic value for following using.

Argument 2: As a basis of our models the probability theory and the theory of regenerative processes (i.e. recurring processes) are used. Proofs of basic theoretical results are received, for example, by [14-15]. If to return in the 70-s of the last century we may remember the boom of mathematical modelling, defining calls flow reliable and time-probabilistic characteristics. The boom passed and appeared the reliability theory, the queuing theory and a variety of models, which proved themselves to be effective. There are created standards and other normative documents regulating system methodical evaluations on the basis of these models. Nowadays these models are widely used and trusted because they produce reliable results confirmed in the course of time. It is worth to remind that these created theories and models are based on the probability theory and the theory of regenerative processes. Some the offered models are the classical models of the 70-s improved and developed to meet the requirements of the present time. The other models [1-12] are created on the basis of the limit theorem for regenerative processes developed in the 70-80-s in Moscow State University.

Argument 3: Skilled analysts know that if a probabilistic analytical model is incorrect then if input data are changed in the range from –¥ to +¥ there are always errors appearing either in infraction the probability theory laws or in illogic of dependencies behavior (most probably on the bounds of possible values) or in impossibility of obtained effects physical explanation. Bounds of input data in the offered software tools are assigned in the range from –¥ to +¥ (more precisely from milliseconds to 10^8 years). Ten-years testing of models including beta testing by different independent companies raise confidence in software tools algorithmic correctness.

Argument 4: As far as possible any designer tends to use several models of different authors. If results of different models use are not divergent a designer begins to trust not only to results but also to the models. Comparison of results of the presented software tools with results of other models use proved their high adequacy (concerning computations of reliability and time-probabilistic characteristics, the other models don't have analogues).

These arguments are supported by correct results of hundreds deep researches and technical solutions corresponding theory and practice.

The offered software tools are an original Russian creation patented. They have been presented at seminars, symposiums, conferences, ISO/IEC working groups and other forums since 2000 in Russia, Australia, CanadaChina, Finland, France, Germany, Kuwait, Luxembourg, Poland, Serbia, Ukraine, the USA, etc. The software tools were awarded by the Golden Medal of the International Innovation and Investment Salon and the International Exhibition "Intellectual Robots", acknowledged on the World's fair of information technologies CeBIT in Germany, noted by diplomas of the Hanover Industrial Exhibition and the Russian exhibitions of software. The offered technology of modelling through the Internet has been acknowledged as the best project-2007 by the National Association of Innovations and Developments of Information Technologies of Russia.

REVIEW AND ANALYSIS OF SYSTEM PROCESSES

As a result of analyzing practice approaches to safety (to industrial, fire, radiating, nuclear, chemical, biological, transport, ecological systems, safety of buildings and constructions, information security) we made the next conclusions-see **Figure 1** [7-12].

For the spheres of industrial, fire, radiating, nuclear, aviation safety in which already there were numerous facts of tragedies-requirements to admissible risks are expressed quantitatively at probability level and qualitative at level of necessary requirements to the initial materials, used resources, protective technologies and operation conditions. Generally risk estimations from one sphere do not use in others spheres because of methods and metrics for risk analysis are different, interpretations are not identical in spite of processes are logically similar.

For the spheres of chemical, biological, transport, ecological safety, safety of buildings and constructions, information security, including the conditions of terrorist threats—requirements to admissible risks are set mainly at qualitative level in the form of requirements to performance. The analytical methods for quantitatively risk analysis are in creating yet. The term "Admissible risk" can't be defined because of one depend on methods. Experience from other spheres is missing.

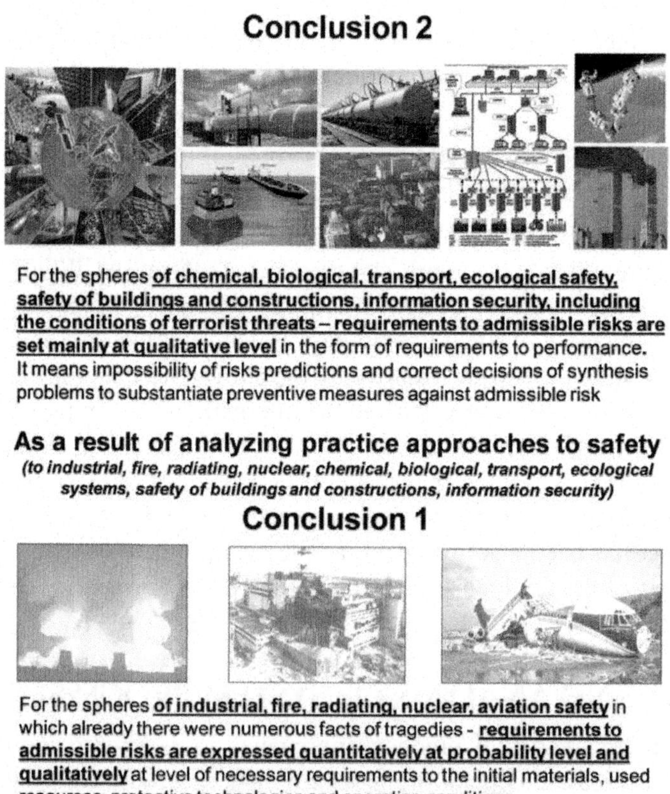

Figure 1: Comparison results to risk estimations.

To improve essentially this situation the offered way includes mathematical models and applicable technologies to predict, analyze and optimize quality and risks for complex systems.

Existing practices for providing system quality and safety were reviewed and analyzed (including approaches of system standards ISO 9001, ISO/IEC 15288, IEC 60300, 61508, CMMI etc.).

As a result of reviewing: all organizations need adequate knowledge to solve the problems, but only some part from them uses modelling complexes; used models are highly specialized, input and calculated metrics are adhered strongly to specificity of systems; existing modelling complexes have been created within the limits of concrete systems and as a rule are very expensive for adaptation to other conditions. In general case prediction and optimization of quality and risks should be founded on the mathematical modelling of system processes. Really, any process is a repeated sequence of consuming time and resources for outcome receiving in all application areas. The moments for any activity beginning and ending are, in mathematical words, random events on time line. Moreover, there exists the general property of all process architectures. It is a repeated performance for majority of timed activities (evaluations, comparisons, selections, controls, analysis etc.) during system life cycle—for example see on Figures 2 and 3 the problems that are due to be and can be solved by the mathematical modelling of processes according to ISO/IEC 15288 "System engineering. Processes of system life cycle". The summary of the analysis is the next. Probability estimations of identical processes from one sphere do not use in other spheres because of methodologies are different, interpretations are not identical. The methods for quantitative quality and risk analysis on probability level are in creating stage yet. As consequence probability estimations are not comparable, experience from different spheres is missing, a universal objective scale of measurement is not established yet. Moreover the terms "Acceptable quality" and "Admissible risk" should be defined on probability scale level only in dependence on corresponding methods.

It does not allow to solve the main problems of a substantiation of system requirements to processes parameters of information gathering and analysis, control, monitoring and counteraction measures at restrictions, and also to confirm about efficiency of the prevent measures to provide quality and safety in different spheres.

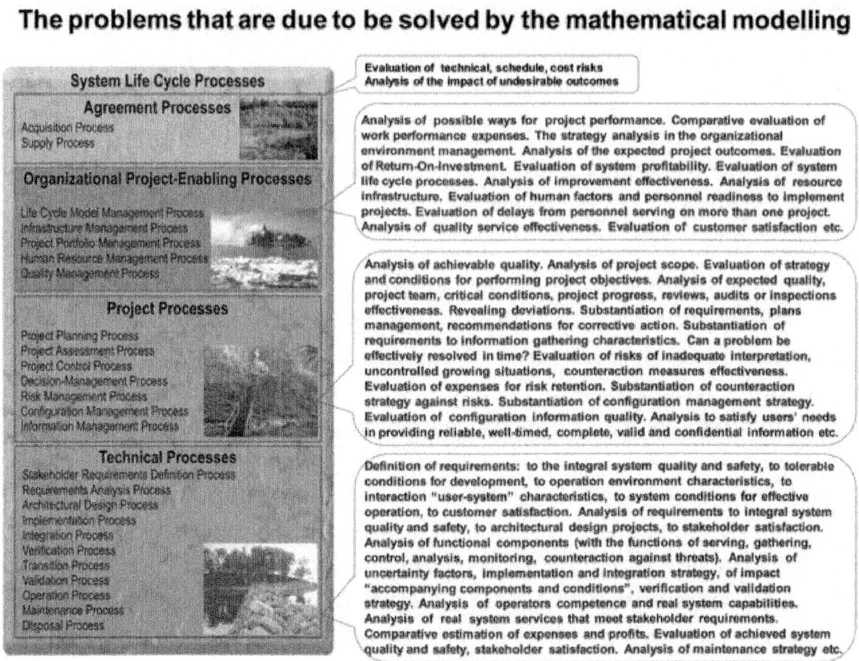

Figure 2: The problems that are due to be and can be solved by mathematical modelling of processes.

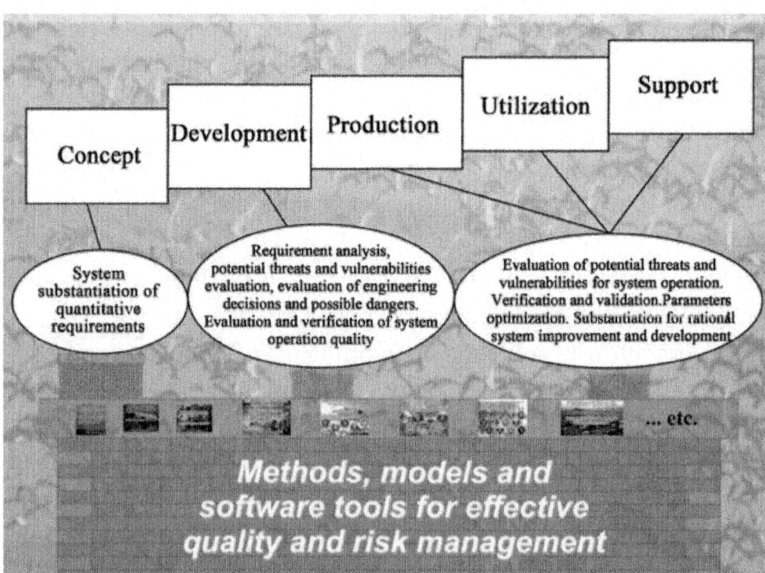

Figure 3: System engineering problems in life stages.

This work focuses on the way for using universal metrics in a system processes: probabilities of success or failure during a given period for an element, subsystem, system. Calculation of these metrics within the limits of the offered probability space built on the basis of the theory for random processes, allows to predict quality and risks on an uniform probability scale, quantitatively to prove comprehensive levels of acceptable quality and admissible risks from "precedents cases". The prediction of risks can use widely transportation safety monitoring data and statistics. In general case a probabilistic space (Ω, B, P) for the evaluation of system operation processes should be proposed, where: Ω —is a limited space of elementary events; B—a class of all subspace of Ω-space, satisfied to the properties of s-algebra; P—a probability measure on a space of elementary events Ω Because, $\Omega = \{w_k\}$ is limited, there is enough to establish a reflection $w_k \to p_k = P(w_k)$ like that $p_k \geq 0$ and $\sum_k p_k = 1$.

EXAMPLE OF MODELLING PROCESSES

The Models and Software Tools to Analyze Information System Processes

The example of creating models and technologies to predict quality and risks is modelling software Complex for Evaluation of Information Systems Operation Quality, patented by Rospatent No. 2000610272 (CEISOQ+) [1- 3].

Requirements to Information Systems (IS) operation depend on SYSTEM purposes and general purpose of IS operation, real conditions (including potential threats), available resources, information sources facilities and communication requirements (see **Figure 4**). This is the logical basis to create universal mathematical models to estimate the reliability and timeliness of information producing, the completeness, validity and confidentiality of the used information from users' point of view [1-3].

The idea of estimating IS operation quality appeared as a result of studying potential threats to output information (see **Figure 5** and example of modelling protection processes against dangerous influences in subsection 3.2).

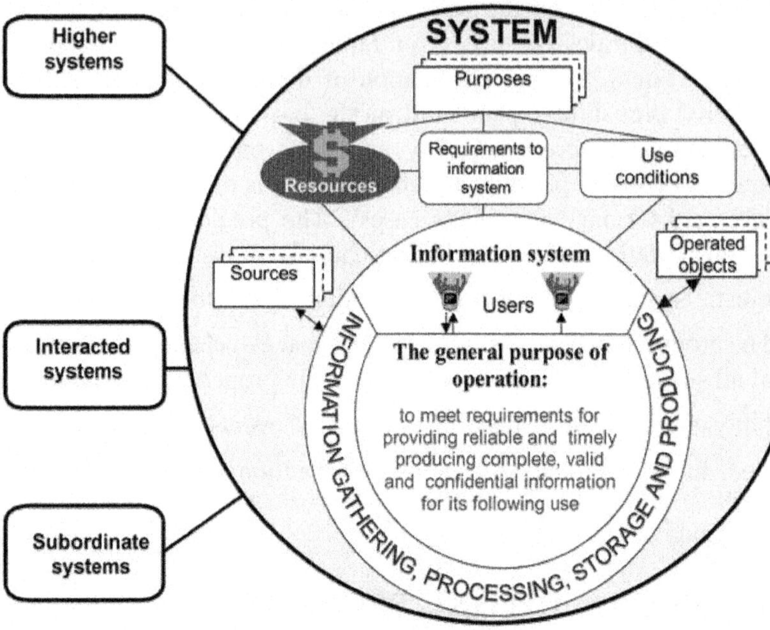

Figure 4: The place and the purpose of information system in a SYSTEM.

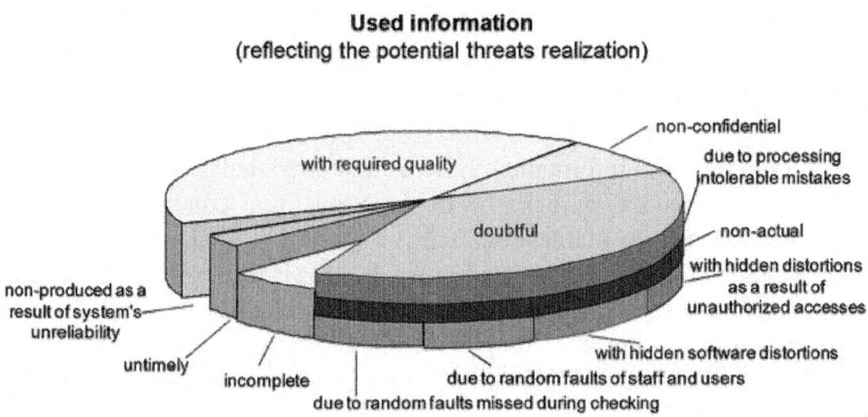

Figure 5: Potential threats to output information according to general purpose of IS operation.

The created CEISOQ+ allows to simplify and to spread the use of the next mathematical models: of functions performance by a system in conditions of unreliability of components; complex of calls processing; of entering into IS current data concerning new objects of application domain; complex of information gathering from sources; of information analysis; of dangerous influences on a protected system; of an unauthorized access to system resources. CEISOQ+ may be applied for solving such system problems appearing in IS life cycle as: substantiation of quantitative system requirements to hardware, software, users, staff, technologies; requirements analysis; estimation of project engineering decisions and possible danger; detection of bottle-necks; investigation of problems concerning potential threats to system operation and information security; testing, verification and validation of IS operation quality; rational optimization of IS technological parameters; substantiation of plans, projects and directions for effective system utilization, improvement and development.

Example of Modelling Protection Processes against Dangerous Influences

Nowadays at system development and utilization an essential part of funds is spent on providing system protection from various dangerous influences able to violate system integrity (these may be failures, defects events, "human factors" events, terrorists attacks, etc.).

There are examined two general technologies of providing protection from dangerous influences in different transportation spheres: proactive diagnostic of system integrity[1] (technology 1) and security monitoring when system integrity is checked at every shift change of operators (technology 2).

Technology 1 is based on proactive diagnostics of system integrity. Diagnostics are carried out periodically. It is assumed that except diagnostics there are also included means of necessary integrity recovery after revealing of danger sources penetration into a system or consequences of negative influences. Integrity violations detecting is possible only as a result of diagnostics, after which system recovery is started. Dangerous influences on system are acted step-by step: at first a danger source penetrates into a system and then after its activation begins to influence. System integrity is not considered to be violated before a penetrated danger source is activated. A danger is considered to be realized only after a danger source has influenced on a system. The essence of protecting process architecture for the first technology is illustrated by **Figure 6**. The cases 1, 4 illustrate dangerous influences. The cases 2, 3, 5 illustrate secure system operation during period T_{req}.[2]

Technology 2, unlike the previous one, implies that operators alternating each other trace system integrity between diagnostics. In case of detecting a danger source an operator is supposed to remove it recovering system integrity (ways of danger sources removing are analogous to the ways of technology 1). A penetration of a danger source is possible only if an operator makes an error but a dangerous influence occurs if the danger is activated before the next diagnostic. Otherwise the source will be detected and neutralized. Errorless operator's actions provide a neutralization of a danger source trying to penetrate into a system. When operators alternate a diagnostic and recovery of lost integrity is held.

For all technologies availability of means of danger sources total-lot detecting and existence of ways of violated system integrity total-lot recovery may seem to be a very high requirement. Nonetheless, a system which can't check and recover its integrity (if it needs) is a very vulnerable and knowingly doomed system.

The probability of safe system operation within the assigned period may be estimated as a result of use the next mathematical models. Risk to lose safety is an addition to 1 (assumption: for all time input characteristic the probability distribution functions (PDF) exist).

There are possible the next variants: variant 1—the assigned period T_{req} is less than established period between neighboring diagnostics $\left(T_{req.} \geq T_{betw.} + T_{diag.}\right) \cdot T_{betw.}$; variant 2— the assigned period $T_{req.}$ is more than or equals to established period between neighboring diagnostics $T_{req} \geq (T_{betw.} + T_{diag.}) \cdot T_{betw..}$—is the time between the end of diagnostic and the beginning of the next diagnostic ($T_{betw.}$ = const); $T_{diag.}$—is the diagnostic time ($T_{betw.}$ = const)· $W_{penetr}(t)$—is the PDF of time between neighboring influences for penetrating a danger source; $W_{activ.}(t)$—is the PDF of activation time of a penetrated danger source; A(t) is the PDF of time between operator's error; $T_{req.}$—is the required period of system operation for prediction.

Statement 1. Under the condition of independence of considered characteristics the probability of dangerous influence absence within the assigned period $T_{req.}$ for the variant 1 of technology 1 is equal to

$$P_{infl.(1)}\left(T_{req.}\right) = 1 - \Omega_{penetr.} * \Omega_{activ.}\left(T_{req.}\right), \tag{1}$$

where *—is the convolution sign.

Statement 2. Under the condition of independence for considered characteristics the probability of dangerous influence absence within the assigned period $T_{req.}$ for the variant 2 of technology 1 is equal to

$$P_{infl.(2)} = \frac{N\left(T_{betw.} + T_{diag.}\right)}{T_{req.}} \cdot P^{N}_{infl.(1)}\left(T_{betw.} + T_{diag.}\right)$$

$$+ \frac{T_{req.} - N\left(T_{betw.} + T_{diag.}\right)}{T_{req.}} P_{infl.}\left(T_{betw.} + T_{diag.}\right),$$

where

$N = \left[T_{\text{req.}} / \left(T_{\text{betw.}} + T_{\text{diag.}}\right)\right]$—is the integer part.

Statement 3. Under the condition of independence for considered characteristics the probability of dangerous influence absence within the assigned period $T_{\text{req.}}$ for the

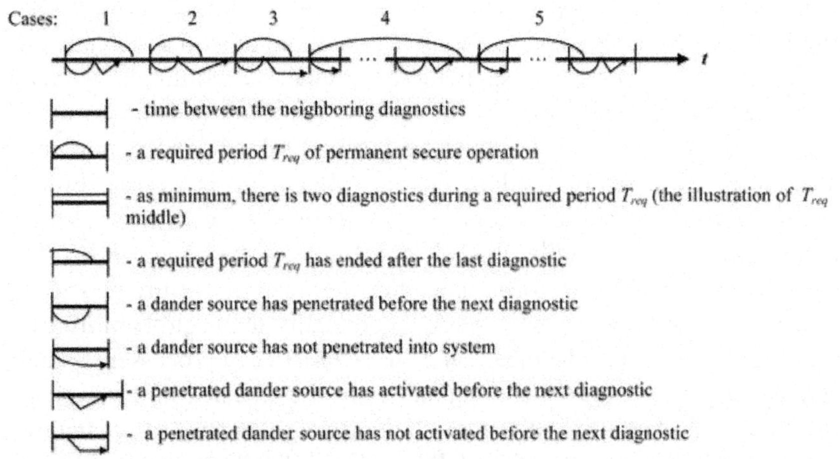

Figure 6: Abstract formalization for technology 1.

variant 1 of technology 2 is equal to

$$P_{\text{inf.(1)}}\left(T_{\text{req.}}\right) = 1 - \int_{0}^{T_{\text{req.}}} \mathrm{d}A(\tau) \int_{0}^{T_{\text{req.}}-\tau} \mathrm{d}\Omega_{\text{penetr.}} * \Omega_{\text{act.}}(\theta). \tag{2}$$

Statement 4. Under the condition of independence of considered characteristics the probability of dangerous influence absence within the assigned period $T_{\text{req.}}$ for the variant 2 of technology 2 is equal to

$$P_{\text{inf.(2)}}\left(T_{\text{req.}}\right) = \frac{N\left(T_{\text{betw.}} + T_{\text{diag.}}\right)}{T_{\text{req.}}} \cdot P_{\text{wholly}}^{N} + \frac{T_{\text{rmn}}}{T_{\text{req.}}} \cdot P_{\text{infl.(1)}}\left(T_{\text{rmn}}\right),$$

P_{wholly}—is the probability of dangerous influence absence within the assigned period $T_{\text{req.}}$, and $P_{\text{infl.(1)}}\left(T_{\text{rmn}}\right)$ is defined above, but one is calculated not for all period $T_{\text{req.}}$, only for the remainder time

$$T_{\text{rmn}} = T_{\text{req.}} - N\left(T_{\text{betw.}} + T_{\text{diag.}}\right).$$

The final clear analytical formulas for modelling are received by convolution of (1) and Lebesque-integration of (2) expression with due regard to Statements 1-4.

The Idea of Modelling Complex Processes

The idea of modelling complex processes consists in the following. Any process represents a set of the works which are carried out with any productivity at limitations for resources and conditions. This amount of works is characterized by expenses of resources (cost, material, human), accordingly works can be executed for different time with various quality. For every system the terms "quality" and "safety" should be formal defined. And conditions are characterized by a set of random factors influencing processes. From the point of view of probability theory and the theory of regenerating processes it is possible to put formally, that all processes on macro-and micro-levels are cyclically repeated. If to assume, that number of recurrences of such processes is very large we can speak theoretically about probability of any events which can occur. The mean time characteristics of processes, frequency characteristics of any events and characteristics, connected in due course are used as input. Probabilities of "success" during a given time of prediction are final or intermediate results of modelling. Risks of failures are an addition to 1. They are used as evaluated output.

Thus the main proposition, implemented in the offered models, concludes the next: all amounts of works, characteristics of their performance, possible events and other inputs are interpreted as expense of time which can be reflected on a timeline. Probability metrics on the introced limited space of elementary events are calculated by the rule of the probability theory.

Correct integration of probability metrics for complex processes are based on a combination and development of models [4-12]. For a complex estimation of the sysms with parallel or consecutive structure existing models can be developed by usual methods of probability theory. For this purpose in analogy with reliability it is necessary to know a mean time between violations of integrity for each of element (similarly mean time between neighboring failures in reliability (MTBF), but in application to violation of quality, safety etc. For unrenowal objects this is mean time to the first failure). Let's consider the elementary structure from two independent series elements that means logic connection "AND" (**Figure 7**, left), or parallel elements that means logic connection "OR" (**Figure 7**, right).

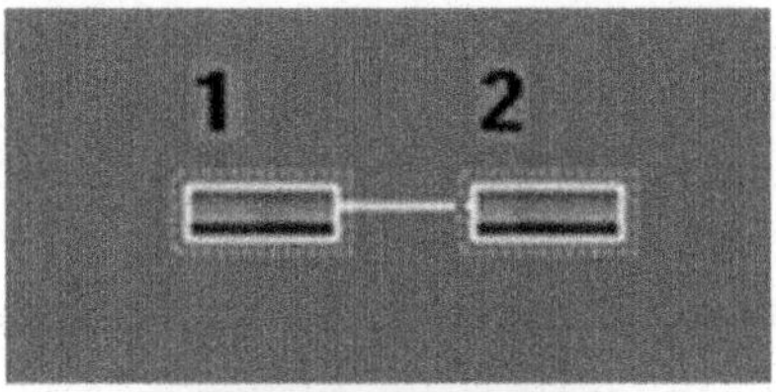

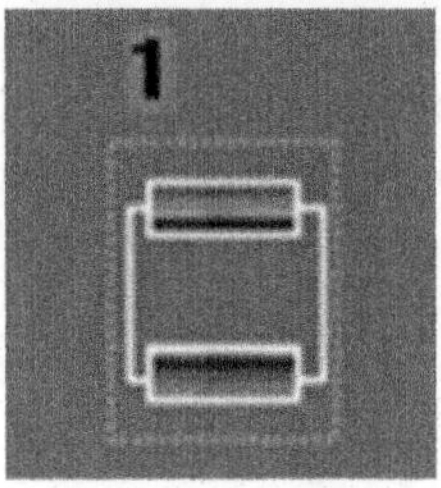

Figure 7: Illustration of system, combined from series (left) or parallel (right) elements

Let's PDF of time between violations of i-th element integrity as $B_i(t)=P(\tau_i \le t)$, then:

1) time between violations of integrity for system comned from consecutively connected independent elents is equal to a minimum from two times τ_i: failure of 1st or 2nd elements (i.e. the system goes into a state of violated integrity when either 1st, or 2nd element integty will be violated). For this case the PDF of time between violations of system integrity is defined by expression

$$B(t) = P(\min(\tau_1,\tau_2)\le t)=1-P(\min(\tau_1,\tau_2)>t) =1-P(\tau_1>t)P(\tau_2>t)=1-[1-B_1(t)][1-B_2(t)]. \quad (3)$$

2) time between violations of integrity for system combined from parallel connected independent elements (hot reservation) is equal to a maximum from two times τ_i: failure of 1st or 2nd elements (i.e. the system goes into a state of violated integrity when both 1st and 2nd element integrity will be violated). For this case the PDF of time between violations of system integrity is defined by expression

$$B(t)=P(\max(\tau_1,\tau_2)\le t) =P(\tau_1\le t)P(\tau_2\le t)=B_1(t)B_2(t) \quad (4)$$

Applying recurrently expressions (3)-(4), it is possible to receive PDF of time between violations of integrity for any complex system with parallel and/ or consecutive structure (in an assumption of independence). The illustration of threats, periodic control, monitoring and recovery of integrity for combined subsystems of estimated system is reflected on **Figure 8**.

Many models are applicable to the system presented as one element. The main output of such system modelling is probability of providing system integrity or violation of system integrity during the given period. If a probability for all points $T_{given.}$ from 0 to ∞ will be calculated, a trajectory of the PDF (or analogy of PDF) for each combined element depending on threats, periodic control, monitoring and recovery of integrity is automatically synthesized. The known kind of this PDF allows to define mean time of providing integrity or between violations of system integrity for every system element by traditional methods of mathematical statistics.

Thus, there is possible an integration of metrics on the level of a PDF of permanent system integrity time or violation of system integrity (or analogy of PDF). And it is the base for quality and risk prediction.

Some Examples of Original Software Tools to Predict Quality and Risks

The next complex for modelling system life cycle processes "MODELLING OF PROCESSES", patented by Rospatent No. 2004610858, supports more than 100 models and includes multi-functional software tools—sees **Figure 9** [7-12].

Complex "MODELLING OF PROCESSES" includes multi-functional software tools for evaluation of Agreement (models and software tools "Acquisition", "Supply"), Enterprise (models and software tools "Environment Management", "Investment Management", "Life Cycle Management", "Resource Management", "Quality Management"), Project (models and software tools "Project Planning", "Project Assessment", "Project Control", "Decision-making", "Risk Management", "Configuration Management", "Information Management") and Technical Processes (models and software tools "Requirements Definition", "Requirements Analysis", "Architectural Design", "Human Factor", "Implementation", "Integration", "Verification", "Transition", "Validation", "Operation", "Maintenance", "Disposal" tools)—see Figures 10-13 (one separate box is an implementation of one or more mathematical models [1-12]).

The one from last implementations is the "Complex for evaluating quality of production processes" (patented by Rospatent No. 2010614145)—**Figure 14**.

The offered models help to answer the system engineering questions (see Figures 2-5) by estimations of quality and risks. The effect from implementation in system life cycle is commensurable with expenses for its creation (see **Figure 15** and www.mathmodels.net).

Thereby necessary attributes of the offered innovative approach to improve system processes are above formed. Traditional approaches consist as a matter of fact in a pragmatical filtration of the information. In the decisions the responsible person, making decision, is guided firstly by the own experience and the knowledge and the advices of those persons of a command to whom trusts. Intuitively forming ideas which seem correct, this person chooses only that information which proves idea.

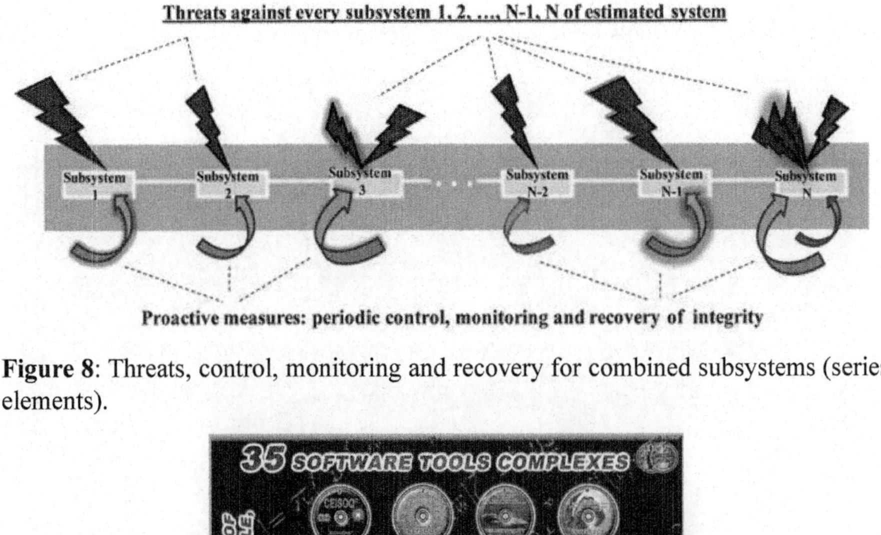

Figure 8: Threats, control, monitoring and recovery for combined subsystems (series elements).

Figure 9: Complexes for modelling system processes.

The denying information is often ignored and more rare— leads to change of initial idea. This approach can be explained from the facts that at absence or limitation of used models it is difficult to investigate at once many ideas for short time. The presented models, methods and software tools, reducing long time of modelling (from several days, weeks and months to few seconds and minutes) change this situation cardinally.

Figure 10: Software tools for evaluation of agreement processes.

Figure 11: Software tools for evaluation of enterprise processes.

The offered innovative approach is at the beginning substantiation of the system requirements, purposefully capable to lead to a success. Further, the responsible person, equipped by a set of necessary mathematical models and their software tools possibilities to predict quality and risks, is powered for generation of the proved ideas and effective decisions. These decisions are physically clear because of using accessible and operative analysis and optimization of processes in system life cycle. The offered approach allows going “from a pragmatically filtration of information to generation of the proved ideas and effective decisions”.

The use of created methods to analyze and optimize system processes allows to optimize quality and risks in practice of system engineering.

OPTIMIZATION OF SYSTEM QUALITY AND RISKS

Classical examples of optimization generally are maximization of a prize (profit, a degree of quality or safety, etc.) at limitations on expenses or minimization of expenses at limitations on an admissible level of quality and/ or safety. In a life cycle of systems criteria and limitations vary.

The statement of problems for system analysis includes definition of conditions, threats and estimation a level of critical measures. As probability parameters give higher guarantees in estimations of a degree of achieving purposes in comparison with average value at a choice it is recommended to use probability as the cores. And evaluated mean time characteristics (for example the mean time between violations of admissible system operation reliability) are auxiliary. For example there are applicable the next general formal statements of problems for system optimization:

Figure 12: Software tools for evaluation of project processes.

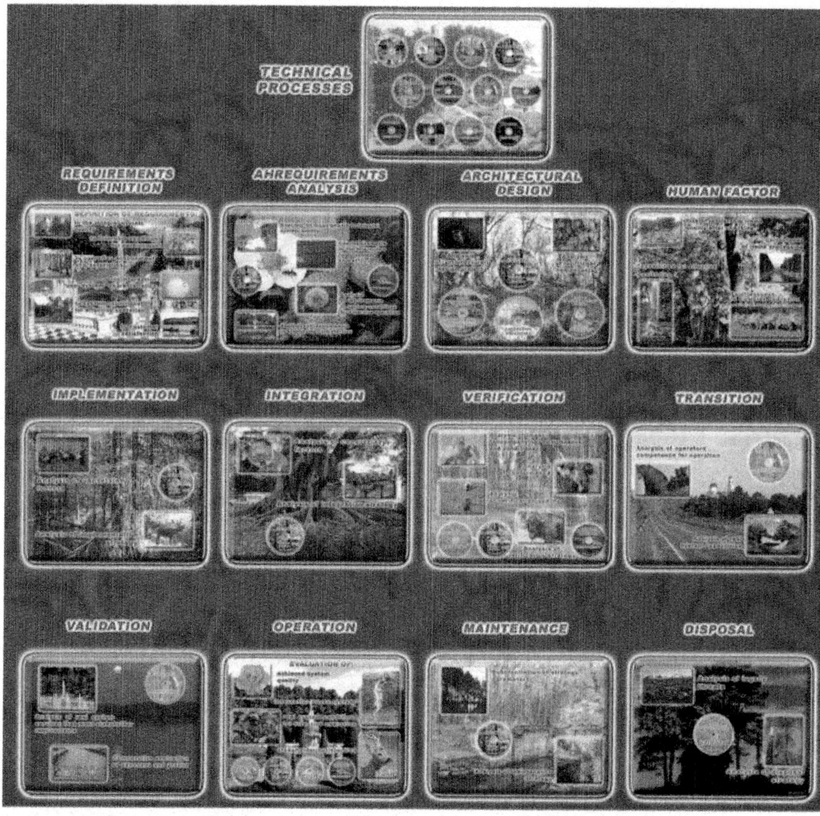

Figure 13: Software tools for evaluation of technical processes.

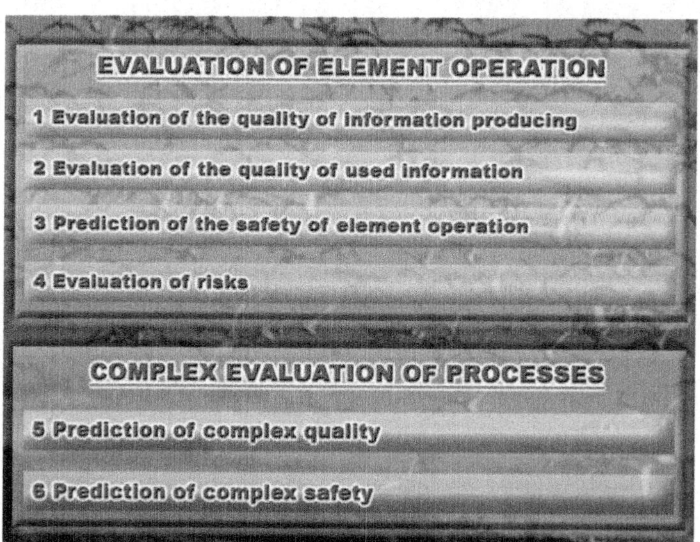

Figure 14: Subsystems of the "complex for evaluating quality of production processes".

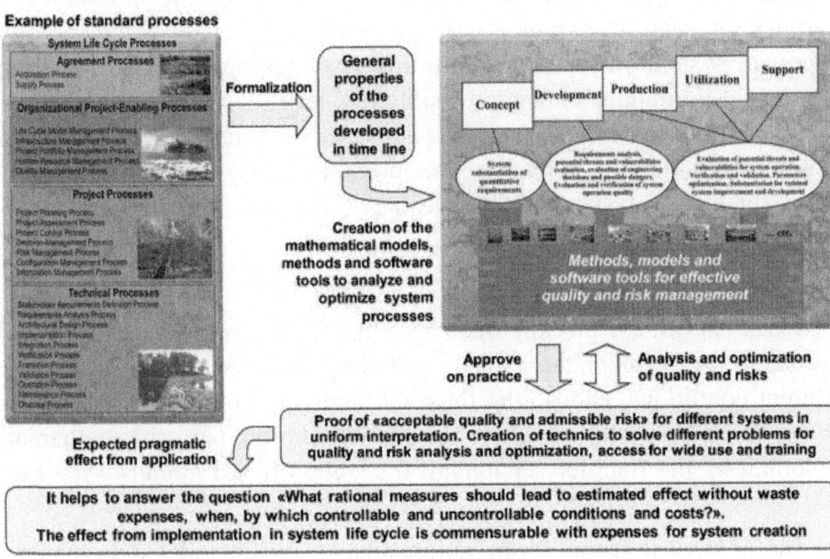

Figure 15: The offered way is the use of created methods to analyze and optimize system processes.

1) on the stages of system concept, development, production and support: system parameters, software, technical and management measures (Q) are the most rational for the given period if on them the minimum of

expenses ($Z_{dev.}$) for creation of system is reached $Z_{dev.}(Q_{rational}) = \min_{Q} Z_{dev.}(Q)$, at limitations on probability of an acceptable level of quality $P_{quality}(Q) \geq P_{adm.}$ and expenses for operation $C_{oper.}(Q) \leq C_{adm.}$ and under other development, operation or maintenance conditions;

2) on operation stage: system parameters, software, technical and management measures (Q) are the most rational for the given period of operation if on them the maximum of probability of providing acceptable system operation quality is reached $P_{quality}(Q_{rational}) = \max_{Q} P_{quality}(Q)$, at limitations on probability of an acceptable level of quality $P_{quality}(Q) \geq P_{adm}$ and expenses for operation $C_{oper.}(Q) \leq C_{adm.}$ and under other operation or maintenance conditions. System parameters, software, technical and management measures (Q) are as rule vectors of input—see examples below.

These statements may be identically transformed into problems of expenses or risk minimization or retention in different limitations. For example for security services it is necessary to provide safety of object, process or system up to the mark. In this case the criterion of a minimum of expenses at limitations on an admissible risk level of dangerous influence on system contrary to counteraction measures or a minimum of risk of dangerous influence at limitations on expenses are possible. There may be combination of formal statements in system life cycle.

The purposed order for use the developed formal approach to analyze and optimize quality and risks is illustrated by **Figure 16**.

When analyst use this approach he'd like for several minutes to formalize a problem, perform mathematical modelling, analyze system processes in different conditions, choose the most rational variant and prepare analytical report. Such possibilities exist: an analyst should perform mathematical modelling by the Internet versions of the some offered models—see **Figure 17**.

The analytical report forms automatically and includes a formalization of analyst's problem, input, results of mathematical modelling in pictures (as demonstrated above in examples), analysis of system processes behavior for different conditions, choice of the most rational variant and recommendations".

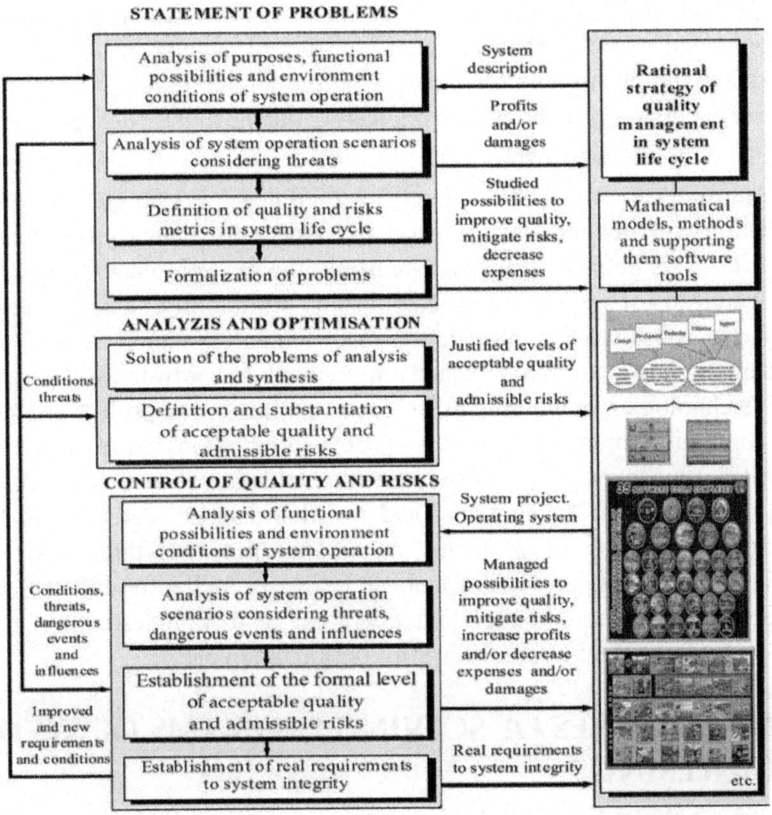

Figure 16: The purposed approach to analyze and optimize quality and risks

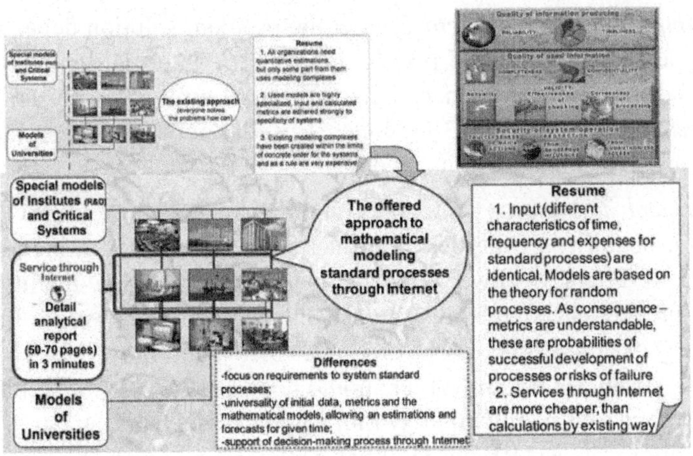

Figure 17: Mathematical modelling by the Internet versions.

It is virtual outsourcing of high system analysis on the base of the offered mathematical models. The purpose is to give to analysts an opportunity of accessible and cheap high technology of studying complex processes in life cycle of estimated systems. This work has begun, the first models are accessible (see www.mathmodels.net).

An application of the offered methodology [1-12 etc.] covers the predictions of probabilities of "success", risks and related profitability and expenses. This helps to solve well-reasonly the next problems in system life cycle: analysis of system use expediency and profitability, selecting a suitable suppliers, substantiation of quality management systems for enterprises, substantiation of quantitative system requirements to hardware, software, users, staff, technologies; requirements analysis, evaluation of project engineering decisions, substantiation of plans, projects and directions for effective system utilization, improvement and development; evaluation of customer satisfaction in system design & development and possible dangers, detection of bottle-necks; investigation of problems concerning potential threats to system operation including protection against terrorists and information security; verification and validation system operation quality, investigation rational conditions for system use and ways for optimization etc.

SOME EXAMPLES OF SOLVING PROBLEMS OF SYSTEM ENGINEERING

Examples 1-9 are presented from simply to complex and based on real input for some operating systems. Example 10 is artificial hypothetic system as a combination of the systems from examples 1-9.

Example 1 ("Human factor"): Let the problem solution depends on joint but independent actions of 5 people. Let each of 4 specialists make 1 error a month and the 5th inexperienced person makes 1 error a day. System recovery time after an error equals to 30 minutes. It is required to evaluate faultlessness of such group's actions within a week.

Approach to solution: Integral computation results by CEISOQ+ reveal that the probability of faultless joint actions of the first 4 skilled specialists within a 40-hours workweek equals to 0.80 but the low-quality work of the 5th unexperienced member mocks the whole group work. Indeed, the probability of faultless actions decreases to 0.15 (see **Figure 18**). The question is lawful—what MTBF an worker should possess to provide a faultlessness of the actions with probability 0.99 within 8 hours of the working day? According to calculations the MTBF not less than 850 working hours is acceptable. It is 106 times (!) more than 8-hours working day.

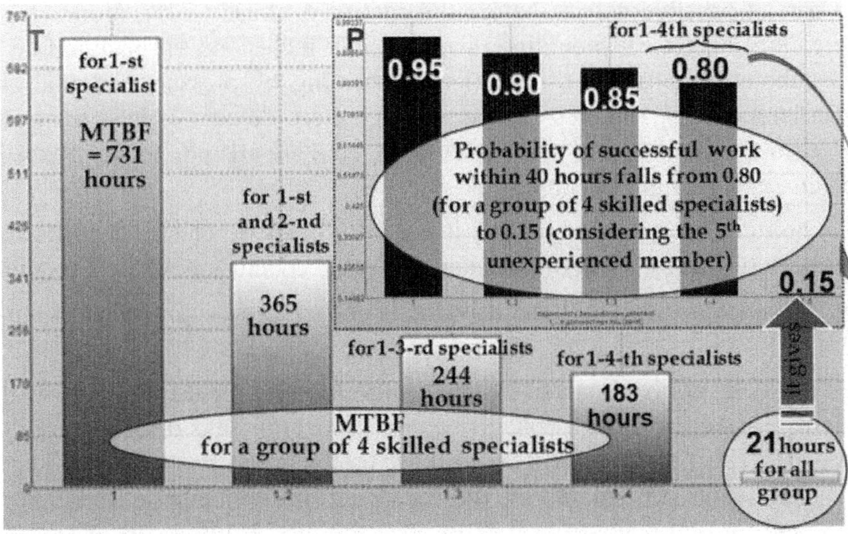

Figure 18: Analysis of human factor.

Example 2 (Efficiency of non-destroying control): Let's consider two competing enterprises which are suppliers of pipes for transportation of production and guided in their quality management system (QMS) by various technical politics. The first of these enterprises, guided by an innovative way of development with rational application of modern information technologies, effectively uses (as believed) existing innovations for quality control. The second company uses cheaper and out-of-date technologies, keeping competitiveness on the market at the expense of it. At the enterprises various methods of non-destroying control are applied to revealing defects.

The first enterprise acquires input production from suppliers after quality control by all recommended methods of non-destroying control (acoustic, magnetic, optical, radiating, radio wave, thermal, electromagnetic etc.) that is confirmed by test reports and certificates on ISO 9001 and on output production. As a result for total controllable production in 100,000 units per a month (for example, production tons, running meters etc.) the part of possible defects before control is 5%, a frequency of errors during the control is no more than 2 defects in a year (these are the latent defects not revealed by existing methods or passed at the control).

The second enterprise is satisfied by certificate on ISO 9001. And only radio wave method of non-destroying control is used by the suppliers. It allows to reveal such defects, as stratifications and deviations on a thickness in metal products (i.e. no more than 10 % of possible defects). At the expense of it the part of possible defects before the control is already 20%, moreover, at the control defects of moulding (slag and flux inclusions, shrinkable bowls, gas bubbles, cracks, etc.), defects of processing by pressure (internal and superficial cracks, ruptures, tempers, dents, etc.), defects of heat treatment (overheats, hardening and hydrogen cracks, etc.) are missed. Totally about 30 defects per a year are possible.

Omitting questions of profits, it needs to compare technical politics of these enterprises by a risk of mistaken analytical conclusion within a month.

Approach to solution: Input and results of control processes are on **Figure 19**.

The comparative analysis of the received dependences has shown: the risk of mistaken analytical conclusions for 1st enterprise is 0.15, and for 2nd one—0.92 (!); if the volume of controllable production is changed from 50,000 to 200,000 units per a month the risk increases for 1st enterprise from 0.08 to 0.58, and for 2nd one—from 0.71 to 0.96; the increase in a part of possible defects twice essentially does not influence value of risk (i.e. efficiency of applied technologies of the control depends essentially on other parameters, in particular from frequency of possible errors); if frequency of possible errors increases twice than the risk increases for 1st enterprise from 0.08 to 0.28, and for 2nd one—from 0.71 to 0.99.

Conclusion: for 1st enterprise the risk of mistaken analytical conclusions at level 0.15 after the control within a month can be recognized as acceptable. The 2nd enterprise supplies frankly defected production (probability nearby 0.9) that will negatively affect further at system operation.

Example 3 (Errors during a use of SCADA system): The control towers use SCADA system (supervisory control and data acquisition) for making decision. Wrong interpretation may be caused by errors of dispatcher personnel, which can miss important information or turn harmless information into dangerous one, fails of SCADA system. Let's consider a control station receiving information from the SCADA system. The information flow is measured in some conventional units and the information flow is of 100 units per hour. The total information contains not more than 1% of data related to potentially dangerous events.

Taking into account automatic data analysis we suppose the speed of event interpretation to be near 30 sec per information unit. In this case 100 information units will be processed during 50 min. At that the frequency of errors for the whole dispatcher shift on duty, including fails of the SCADA system itself is about 1 error per year according to statistical data. The task is to estimate the risk of mistaken analyticcal conclusion for a time period of 1 hour, during one dispatcher shift turn of 8 hours, 1 month, 1 and 10 years.

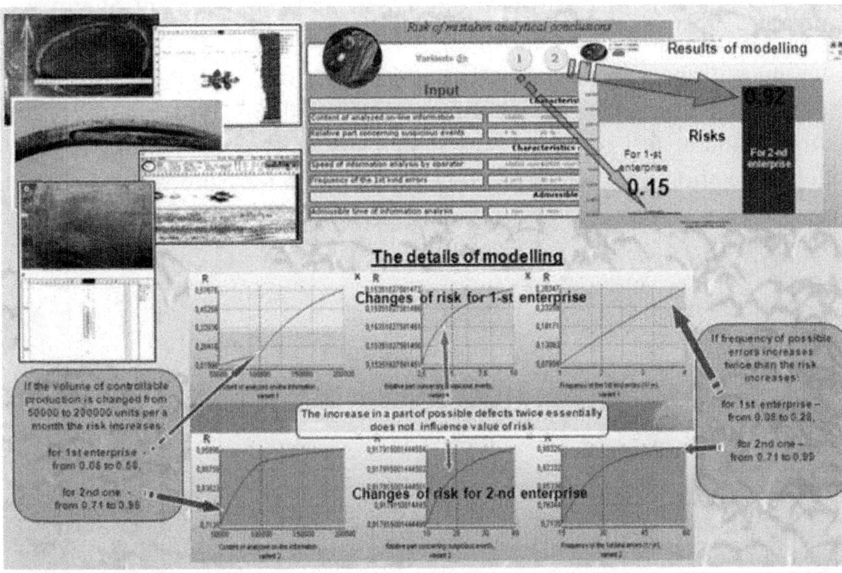

Figure 19: Comparative estimation of efficiency of non-destroying control.

Approach to solution: The analysis of modelling by the software tools “Complex for evaluating quality of production processes” shows (see **Figure 20**) that for short time periods such as one shift turn or even for a month the risk of mistaken analytical conclusion is small enough (0.00076 and 0.07 accordingly). But when the time period grows the risk increases and becomes 0.565 for a year and almost unity (0.9998) during time period of 10 years. This means that during a month the probability for errors of dispatcher personal or SCADA system fails to occur is very small and their operation will be almost faultless. But for a more long time period such as a year is considered 1 - 2 errors of dispatcher personal or system SCADA fails will occur for certain.

Considering high reliability of SCADA system and according to “precedent” principle the level 0.07 for the risk of mistaken analytical conclusion during a month can be defined as admissible.

Example 4 (Efficiency of counteraction measures against risks in pipes manufacture and use): It needs to compare efficiency of counteraction measures against risks for two different companies that are responsible for systems of pipes manufacture and use.

Approach to solution: A solution can be based on comparisons of risks to lose efficiency during 2 years and 15 years of companies operation.

Let's the 1st system is characterized by measures: 1st measure—QMS at the supplier; 2nd measure—production quality check by all recommended kinds and methods of control within a year and improvement of times in 3 years; 3rd measure—the control by SCADA-system; 4th measure—remote sounding with preservation of efficiency within the days, carried out once a week; 5th measure—annual local inspections with preservation of efficiency within a month; 6th measure—integrated inspections of 1 times in 5 years with preservation of efficiency within a month; 7th measure—electrochemical measure—electrochemical protection of pipelines and means of telemechanics.

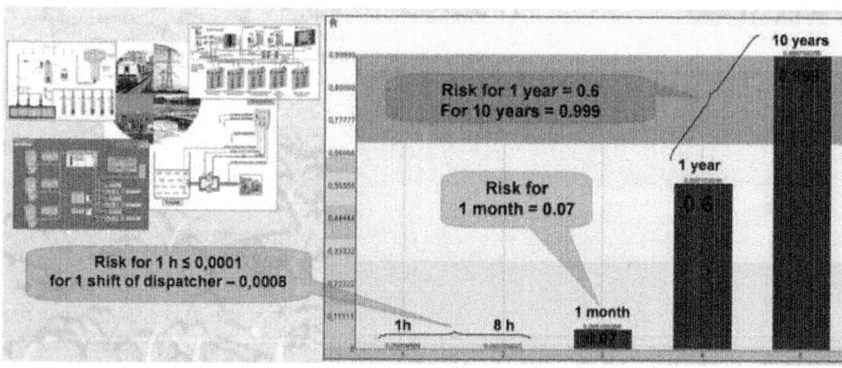

Figure 20: Results of modelling a SCADA-system.

Let's the 2nd system is characterized by measures: 1st measure—QMS at the supplier; 2nd measure-the control by SCADA-system; 3rd measure—helicopter inspection and regular radiographic methods of the analysis with preservation of efficiency within the days, carried out once a week; 4th measure-annual local inspections with preservation of efficiency within a month; 5th measureintegrated inspections of 1 times in 5 years with presservation of efficiency within a month; 6th measure—electrochemical protection of pipelines and means of telemechanics.

Results show high degree of efficiency for both companies: 0.11 - 0.25 during 2 years, 0.21 - 0.38 during 15 years. These results, compared with results from other spheres [7-12] considering "precedent cases" (see also examples of this paper), proves: the level of risks 0.11 for 2 years and 0.25 for the period 15 years can be recognized as "admissible".

Example 5 (Preservation of foods quality): Prediction and optimization of system quality is demonstrated on an example of modelling processes that are peculiar for grain storage. Quality of the grain supplied on longtime storage, decreases because of influences of dangerous biological, chemical and physical factors. Let's estimate the possible period before such moment of time when storing grain begin to loss required quality, and also expediency of introduction of continuous monitoring of grain quality.

Approach to solution is based on the use of the subsystem "Risk evaluation. Risk of uncontrollable development of situations" of the software tools "Complex for evaluating quality of production processes". The list of dangerous factors (threats), controllable parameters and proactive actions at grain storage in real conditions is resulted in **Table 1** [19].

The cleared, dry and non-contaminated grain may be stored lost-free some years. However, the insects which are present in granaries and round them, occupy grain and breed. For example, every 2 months rice weevil increases in the number at 15 - 45 times at temperature from 20°C to 25°C. If in batch of wheat in weight 1000 tons contamination reaches 16 bugs on 1 kg of grain, losses are expected more than 5 %. The grain polluted by wreckers and products of their vital functions (excrements, dead bodies, uric acid, etc.), becomes toxic. It cannot be used for the food purposes. Therefore we will consider security of grain from insects, believing within the example, that exactly the main dangers are from them.

Let's a frequency of latent occurrence of critical situations during hot months is often not less than 1 time a day (i.e. every day at air temperature above 12°C infection or the further damage of grain is possible). Our consideration: at 12°C - 15°C a duration of insects development (for example, weevil) is 141 - 376 days, and in a laying from 300 to 600 eggs a cycle of development is 1.5 - 2 months. In the conditions of cooling of grain below a temperature threshold of insects development (more low than 10.2°C) their pairing, eggs putting off and development of all stages stop. Insects become inactive and almost do not eat. Long stay of insects at such temperature leads to their slow extinction. Besides, humidity maintenance at a level of 13% - 15% also promotes extinction of insects.

Thus, input for modelling is defined: frequency of latent occurrence of critical situations—from 1 time a day to 1 time a week; mean time of danger source activation—1.5 months; time between diagnostics of system integrity (analysis of temperature and humidity)— 1 hour; duration of diagnostic, including recovery time— 1 hour.

Table 1: The list of dangerous factors, controllable parameters and proactive measures at grain storage

Dangerous factors (threats)	Controllable parameters	Proactive measures
Biological: - microorganisms; -contamination of grain stocks by insects	Grain, spoilt as a result of self-warming and growing mouldy. Insects and pincers, a dung of rodents.	Observance of requirements of the standard documentation on grain storage. Complex of practical and exterminating measures against insects.
Chemical: - mycotoxins; -products of fats oxidation in grain (free fat acids, aldehydes, ketones, peroxides); -harmful products of vital functions of grain wreckers; - pesticides	The content of the spoilt and damaged grains as a result of microbiological spoiling. Organoleptic indicators (colour, a smell), and also the content of the beaten and brought down grains. Total density of pollution by live and dead wreckers, no more than 15 copies /kg. Residual quantities.	Observance of the general sanitary norms. Observance of regulations for pesticides use and terms of grain endurance after processing. Decrease of storage temperature to low positive temperatures of air. Observance of the instruction for pest control. Observance of requirements to grain after desinsection.
Physical: -extraneous subjects, casual and weed impurity; -grain temperature and humidity	Rough, large and casual impurity. Stable temperatureand humidity	Grain clearing on separators. Regular cooling of grain to low positive temperature (no more 10°C). Observance of the requirements of the general technological regulations

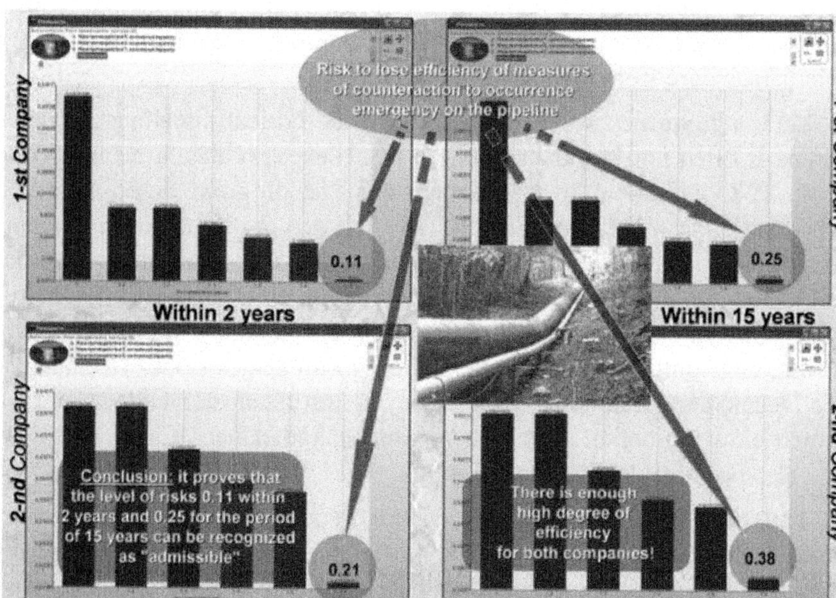

Figure 21: Comparisons of risks to lose efficiency.

It is enough to predict a risk of uncontrollable development of situations with grain storage. The results of modelling for the period from 1 year to 6 years have shown the following.

If a frequency of latent occurrence of critical situations is 1 - 2 times a day, risk of uncontrollable development of situations within a year will grow from 0.28 to 0.47, and during 2-years period it can exceed 0.5—see **Figure 22** left.

These results can be interpreted so: if storage conditions daily promote occurrence of insects, then for a 1 - 2 years grain quality loss is possible at the same degree as preservation of quality. Thus the next conclusion is right: the accepted conditions of grain storage in a granary leads to inadmissible damages. For prevention such danger scenario the following basic requirements [19] should be performed: a smell unusual for grain should not be felt; isolation from dampness and from penetration of subsoil waters should be provided; grain-elevator should not have unfixed vertical and horizontal joints; doors should be densely closed, floors and walls should be smooth, without cracks, roofs—in a serviceable condition; fixtures should be protected by protective caps with grids; inlet of active ventilation should be densely closed preventing a penetration of an atmospheric precipitation, etc.

Performance of these requirements conducts to decrease a frequency of latent occurrence of critical situations in granaries. Further we will answer the question-what about risk in conditions of more rare occurrences of critical situations? And, on the contrary, what the level of a frequency of latent occurrence of critical situations can be considered as admissible for granaries?

Results of modelling show: if frequency of latent occurrence of critical situations will be 1 - 2 times a week, risk of uncontrollable development of situations within a year will grow from 0.05 to 0.09, i.e. the risk decreases in 5 - 7 times! (Against the level from 0.28 to 0.47), and within 6 years risk will make 0.25 - 0.43 (it is better, than risk within a year when frequency of latent occurrence of critical situations is 1 - 2 times a day!)—see **Figure 22** right. These results can be interpreted so: if storage conditions prevent from occurrence of insects with the frequency more often, than once a week, probability of preservation of grain quality within 3 - 6 years exceeds probability of quality loss in 3 - 5 times!

The results of modelling are quantitatively confirmed by results of long-term researches of the Russian Research Institute of Grain [19]. According to these researches experimental batches of grain wheat met to standard requirements of class grain has been kept within 6 years without deterioration in dry, cleared and the cooled condition. Moreover, the received values of risk can define admissible quality for grain storage. Indeed, new recommended result is: the admissble risk of uncontrollable development of situations should not exceed 0.10 for 1 year and 0.25 for 6 years of grain storage. It is comparable with the results of example 4 concerning other sphere of models applications.

***Example 6 (Estimation of control and monitoring for*:** railroad tracks). Geological, ecological, technical, mechanical, information and other factors can impact on system operation. From a general point of view these factors can be divided in the following groups: pressure on the railroad tracks from surrounding soils, lateral earth movements; soil erosion and vegetation impact (for example, in jungle area); thermal effects; slope instability of the railroad tracks right-of-way; defects or deformations in the railroad tracks; the failures of integration; mechanical bumps etc.

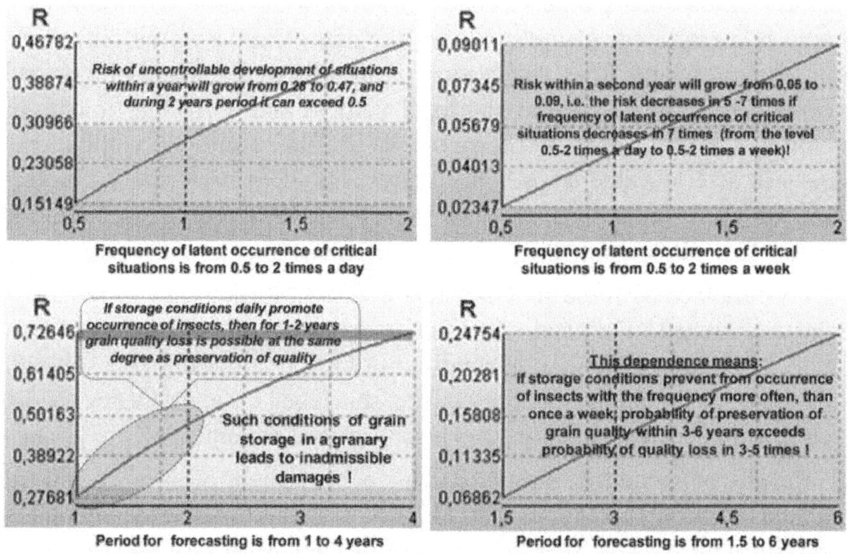

Figure 22: Some detail results of modelling and analysing.

After an analysis of the accident statistics we can assume the main characteristics to have following values: the frequency of critical situations is 3 events per year, the mean time of situation evolution before damaging is 1 hour. The railroad tracks integrity is confirmed on the central control station once in a day while the dispatcher shifts are changed. The system integrity control includes the monitoring of information from SCADA system and other automatic facilities such as data from leak detection integrated system, intelligent electronic devices and others. Duration of integrity control is 1 hour on average. A dispatcher can misunderstand the results of monitoring of the pipeline condition and do not start actions to control critical situation opportunely as required. Taking into account dispatcher personnel training degree and automatic decision support on the statistics we assume the mean time between mistakes for the shift of monitoring to be 1 week or more. The task is to estimate the risk of uncontrolled situation evolution for a time period of 1 month, 1 year, and 10 years in give conditions.

Approach to solution: Input data are determined for evaluation the risk during a time period of 1 month (columns 1, 4), 1 year (columns 2, 5), and 10 years (columns 3, 6); for easy recoverable critical situations with time period of integrity control and recovery of 1 hour (columns 1-3) and for severe critical situations with time period of integrity control and recovery of 10 days (columns 4-6). Results of analysis see on **Figure 23**.

The risk of uncontrolled situation evolution is high enough (more than 0.6 during a year). To decrease the risk the mean time between mistakes for the dispatcher personnel should be increased, the time of carrying out control and repairing damages should be shorten to several days or even hours.

Example 7 (Reliability of engineering equipment for enterprise objects): Let prediction of operation reliability of computer-aided engineering equipment against usual non-automated engineering equipment is needed for the stages "Concept" and "Development". An estimated object (for instance, the center of information processing and storage) includes power supply subsystem, an air conditioning subsystem, supported by 2 sources of an uninterrupted supply and a server, supported by 1 source of an uninterrupted supply and disks for information storage, supported also by 2 sources of an uninterrupted supply. In turn, the power supply subsystem includes the switchboards, supporting by 2 sources of an uninterrupted supply. All listed above engineering equipment is supported by 2 engine-generating installations.

Approach to solution: Within the example two subsystems are allocated (see **Figure 24**): subsystem 1—the city power supply formalized as basic and reserve subsystems; subsystem 2—an object fragment. It is supposed, that reliability of the object operation during given period is provided, if "AND" in 1st subsystem "AND" in 2nd subsystem there will be no power supply infringements.

The analysis of modelling shows, that, at estimated technology of the control, monitoring and integrity recovery the MTBF for computer-aided engineering equipment will equal to 42219 hours. The probability of reliable object operation within a year equals to 0.828. In turn, for usual non-automated engineering equipment (there is no the monitoring implemented for computeraided engineering equipment) efficiency characterized by estimations on **Figure 25**.

For usual non-automated engineering equipment the MTBF will make 16,196 hours (it is at 2.44 time less, than for computer-aided engineering equipment that uses monitoring), and the probability of reliable object operation within a year equals to 0.649 (at 1.26 time less, than for computer-aided engineering equipment). Moreover, without automation for 2 years the probability of at least one failure (0.52) exceeds probability of reliable operation (0.48). Against this the probability of reliable object operation within 2 years for computer-aided engineering equipment is more at 1.5 times and will not fall low than 0.7. Attention, please, results are comparable with the results of examples 4-6 concerning other spheres of models applications.

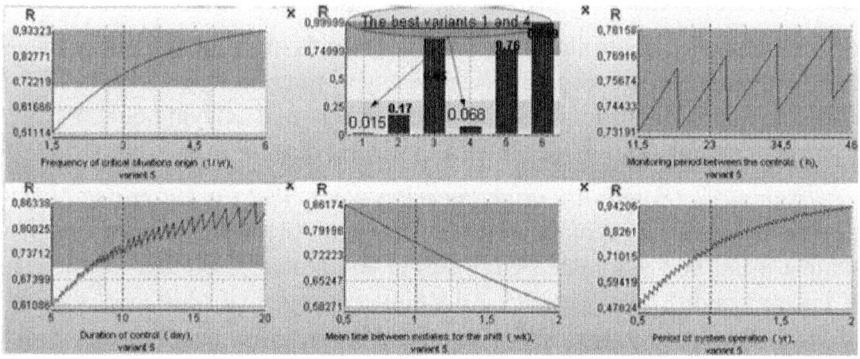

Figure 23: Dependency of the risk for 1 year as input data varying in the range of -50% +100% (variant 5: period of integrity control and recovery =10 days).

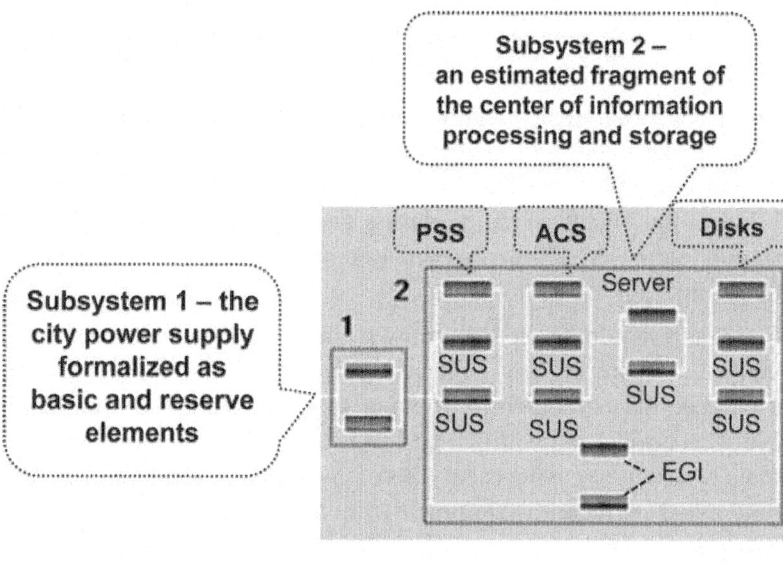

Figure 24: Logic model (PSS—power supply subsystem, ACS—air conditioning subsystem, SUS—source of an uninterrupted supply, EGI—engine-generating installation).

Example 8 (the estimations of flights safety before and after 09/11) [3-4,10-11]. From the modelling point of view a flying airplane is a protected system operating in conditions of threats to its integrity during the flight. What about risk to lose complex safety before and after 09/11? And what about efficiency of additional measures for counteractions?

Approach to solution: For the existing before 09/11 safety system consisted the next main barriers: pass and inter-object modes in aerodromes and centers of air traffic control; preflight examination and control of passengers and their luggage during the registration; preflight examination before boarding; a lock-up door to the cockpit; an on-line warning about a highjacking (this barrier is critical if terrorists try to hide the fact of highjacking).

The results of modelling are the next: before 09/11 in Russia and the USA the risk to lose complex safety against trained terrorists estimated about 0.47 - 0.48, i.e. every second prepared terrorist act comes true. The bottlenecks were a weak protection of a cockpit and absence of active opposing measures on board an airplane (see **Figure 26**).

How the level of the safety may be increased by measures, listed on **Figure 27**? As in a cabin may be accomplices able to repeat the high-jacking after an additional preparing there must be provided ways of compulsory keeping of suspicious passengers on their seats till the emergency landing. All the listed measures seem to be effective but how effective are they quantitatively? It is impossible to make a variety of natural experiments. We use the offered mathematical models.

Analysis of modelling results has shown, that after implementation of the described measures the integrated risk to lose complex safety of flight during 5 hours of flight against terrorist threats is equal to 0.000004. And if duration of threats will be increased to 5 days the risk raises from 0.000004 to 0.002. The last can be commented by the next interpretation: safety will be achieved in 998 cases from thousand hypothetical terrorist attacks. Even taking into account an essential error of initial scenarios and preconditions it is an obvious indicator of high efficiency of additional safety measures according to "precedent" principle! Still it is not a victory. It is clear that the first failures will make terrorists to analyze their causes and find new bottlenecks of the safety system thus continuing the counteraction. This counteraction will be ended when there are taken proactive measures which effectiveness is based on modeling.

Example 9 (Protection against an unauthorized access): We will consider the approach to an estimation of IS protection against an unauthorized access (UAA) and information confidentiality. A resources protection from UAA is a sequence of barriers. If a violator overcomes these barriers he gets access to IS information and/or software resources. In the **Table 2** there are shown supposed characteristics of barriers and mean time of their overcoming by a specially trained violator (real values of such characteristics may be drawn as a result of actual tests or use of other models). It is required to estimate IS protection against UAA.

Approach to solution: The analysis of computed dependencies (see **Figure 28** left) shows the next. The barriers 1-3 will be overcome with the probability equal to 0.63. However, monthly password changing for barriers 4-6 allows to increase the protection probability from 0.37 to 0.94 but the level of IS protection (the first six barriers) is still low. The introducing of 7-9 barriers is useless because it does not practically increase the level of IS protection. The use of cryptography allows to increase the level of IS protection to 0.999. This is probability for all time of IS operation (i.e. about 20 - 30 years). It is possible to establish a conclusion, that with the use of cryptographic devices the achieved protection level exceeds similar level of reliability and safety for processes from examples above. But according to "precedent" principle this level of protection can't be recommended as high for every cases.

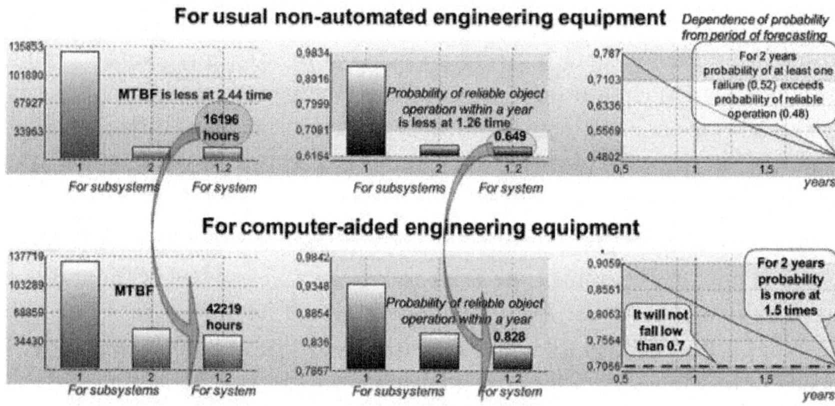

Figure 25: Results of modelling engineering equipment.

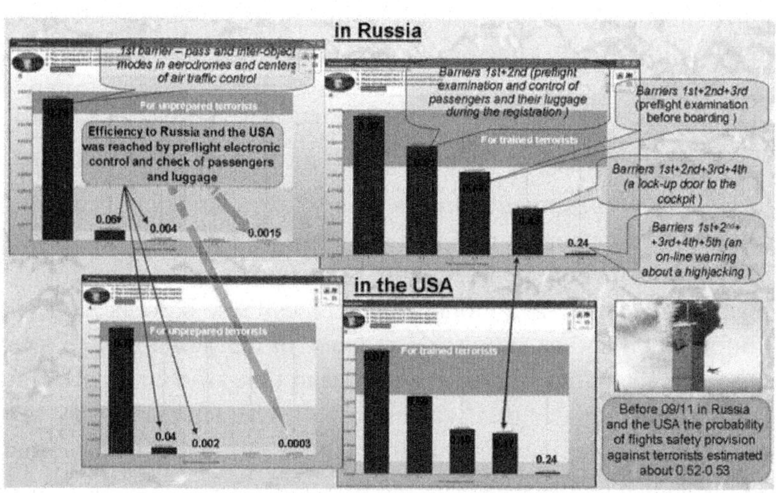

Figure 26: Risk to lose complex safety against terrorists.

Let's look on example condition more widely. The violator is interested in certain IS resources during a given period of time. This period is called the period of objective confidentiality. Let's information confidentiality should be provided within 7 days. **Figure 28** (right) shows how this period influences on protection: in comparison with the results above the use of the first 5 barriers provides confidentiality during 7 days on the level 0.98 which is more higher than protection by the 9 barriers (0.946—see **Figure 28** left); the use of all the 10 barriers provides the required confidentiality on the level 0.99997. It eliminates the customer's risk in providing system protection. It explains the role of a considered period of objective confidentiality. Its consideration allows to understand, that real protection of resources during 7 days is essentially higher—0.99997 against 0.999!

Table 2: Input for modeling.

Barrier	The frequency of barrier parameter value changes	The mean time of the barrier overcoming	Possible way of the barrier overcoming
1. Guarded territory	Every 2 hours	30 min.	Unespied penetration on the territory
2. Admission system for coming into office	Once a day	10 min.	Documents forgery, fraud
3. Electronic key for powering the computer	Every 5 years (MTBF = 5 years)	1 week	Theft, collusion, forced confiscating
4. Password to login	Once a month	1 month	Collusion, forced extortion, spying, password decoding
5. Password for access to devices	Once a month	10 days	Collusion, forced extortion, spying, password decoding
6. Password for requesting information resources	Once a month	10 days	Collusion, forced extortion, spying, password decoding
7. Registered device for information recording	Once a year	1 day	Theft, collusion, forced confiscating
8. Confirmation of user authenticity during a computer session	Once a month	1 day	Collusion, forced extortion, spying
9. Television monitoring	Once a 5 years (MTBF = 5 years)	2 days	Collusion, disrepair imitation, force roller
10. Cryptosystem	1 key a month	2 years	Collusion, deciphering

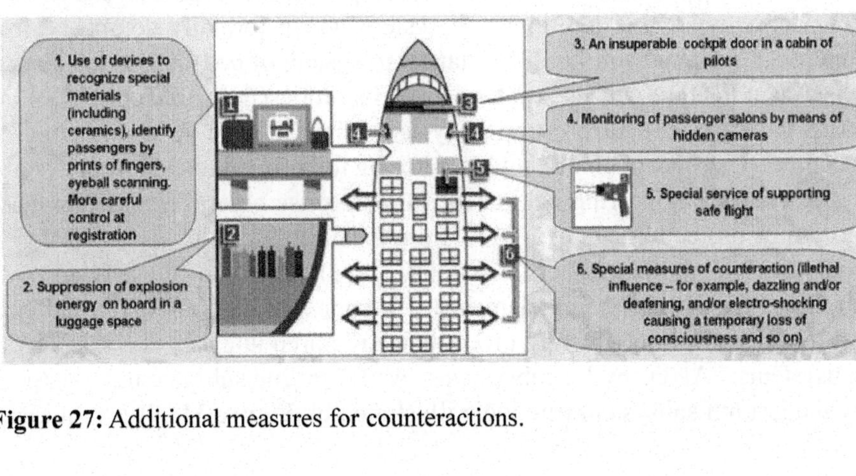

Figure 27: Additional measures for counteractions.

And what about safety of complex system, including head subsystem and two used subsystems 2 and 3 (see **Figure 29**)? A frequency of threats is no more than 1 time at hour, average time for system recovery is no more than 30 minutes. It is required to predict quantitatively the level of safety within month and year system operation and to reveal its bottlenecks.

Results of modelling are reflected on **Figure 30**. With monitoring and control within a month all barriers are overcame with probability 0.9, and within a year—with probability 0.43. Without monitoring and control these probabilities decrease to level 0.83 and 0.29 accordingly. Monitoring is ineffective for example conditions.

Following recommendations are obvious: for safe system operation it is expedient, that all subsystems are strong equally. The technology of safety maintenance in emergency case is necessary. Recommendations are supported by the offered mathematical models.

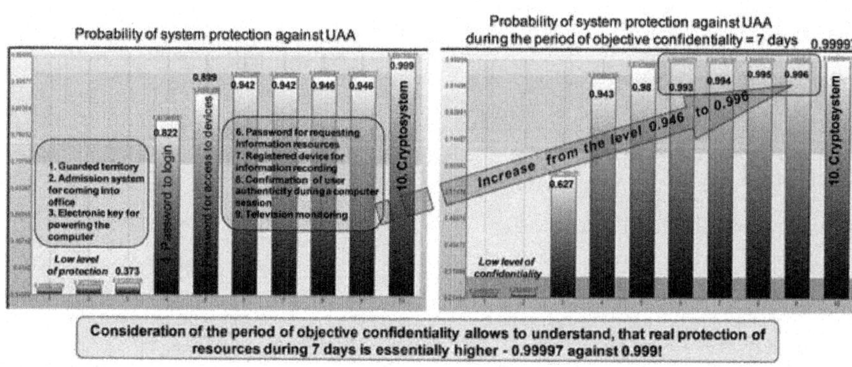

Figure 28: Comparison of protection levels.

Example 10 (Predicts of risks for complex multipurpose system). Let's consider a hypothetic multipurpose system which formally composed from functional system—similar, for instance, to system of non-destroying control, pipes manufacture and use or foods presservation (from examples 2, 4 and 5), gathering and data processing system (similar to SCADA-system from example 3), system of control and monitoring for railroad tracks (from example 7), system of engineering equipment for enterprise object (from example 7), system of protection against UAA (from example 9). "The human factor" is considered in the parameters of control, monitoring and integrity recovery measures for corresponding elements. It is supposed, that a required integrity of system is not lost, if during given time a required integrity is not lost by all subsystems: "AND" by 1st subsystem, "AND" by 2nd subsystem, ×××"AND" by the last 6th subsystem (the logic illustrated by **Figure 31**). It is required to predict risk to lose integrity during years of system operation and estimate the measures of risk management, including the periodic control and, where it is possible, continuous monitoring of integrity of components.

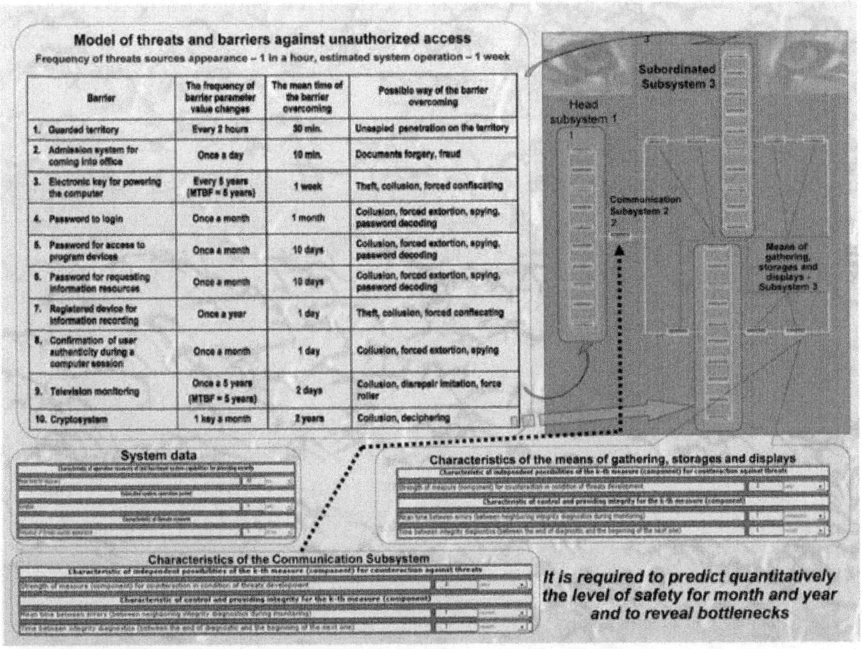

Figure 29: Input for modelling complex safety.

Approach to solution: The input for subsystems 1-6 is described in examples 2-7, 9. The general results of complex prediction of risk are reflected by **Figure 32**. Analysis of results shows, that integrated risk to lose system integrity of system within operational 2 years is 0.27. And for subsystems 1-6 this risk differs from 0.01 to 0.12.

The dependence of integrated risk on time of predicttion (from 1 to 4 years) is reflected by **Figure 33**. Analysis of results shows, that the integrated risk to lose system integrity is changing from 0.11 to 0.67 (with using of measures of the periodic control and where it is possible, monitoring of elements operation).

The general logic proposition is right for a given period of prediction: as a rule, the risk to lose system integrity increases in depending on increasing time period. But there are the features demanding a logic explanation. Serrated and nonmonotonic character of dependence on **Figure 33**is explained by the periodic diagnostics of elements, monitoring presence or absence and their quantitative values (see subsection 3.2). Let's remind: for every monitored element a penetration of a danger source and its activation is possible only if an operator-monitor makes an error but a dangerous influence occurs if the danger is activated before the next diagnostic. Otherwise the source will be detected and neutralized. Immediately after element diagnostic the risk decreases because during diagnostic all dangers are detected and neutralized and at the beginning of a period after diagnostic dangerous influences don't have enough time to accumulate and be activated.

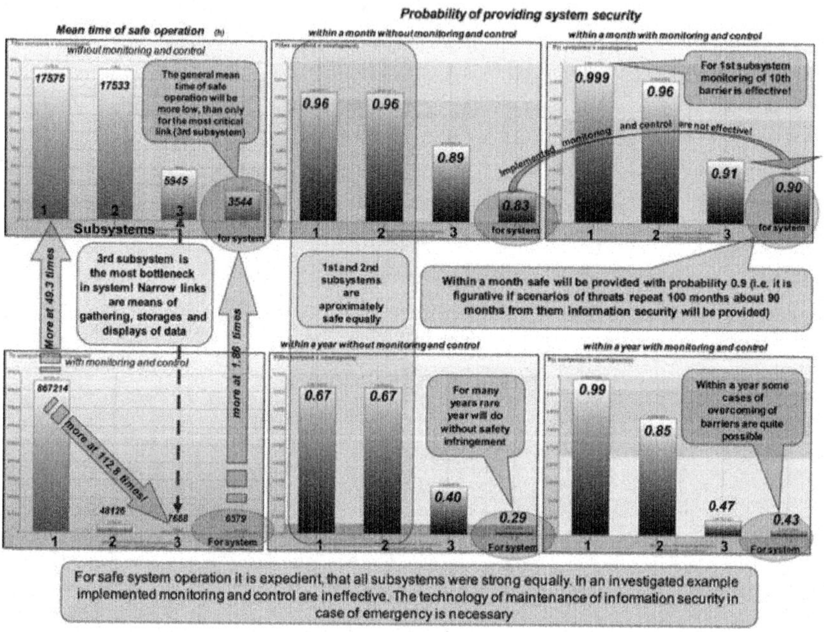

Figure 30: The results of prediction and analysis.

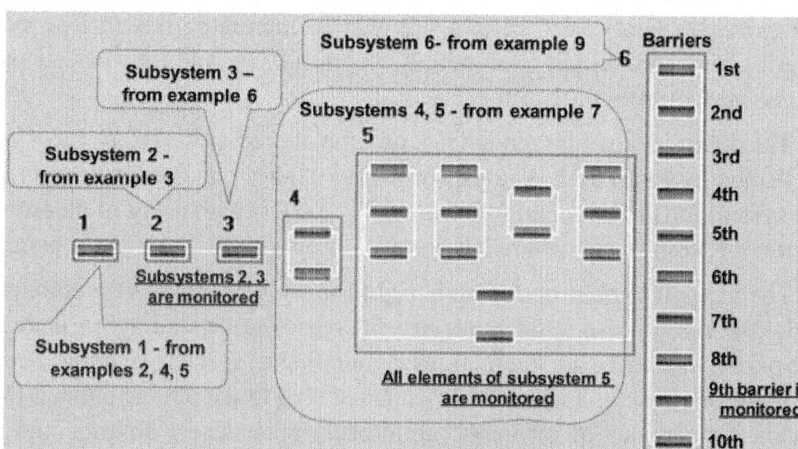

Figure 31: The results of prediction and analysis.

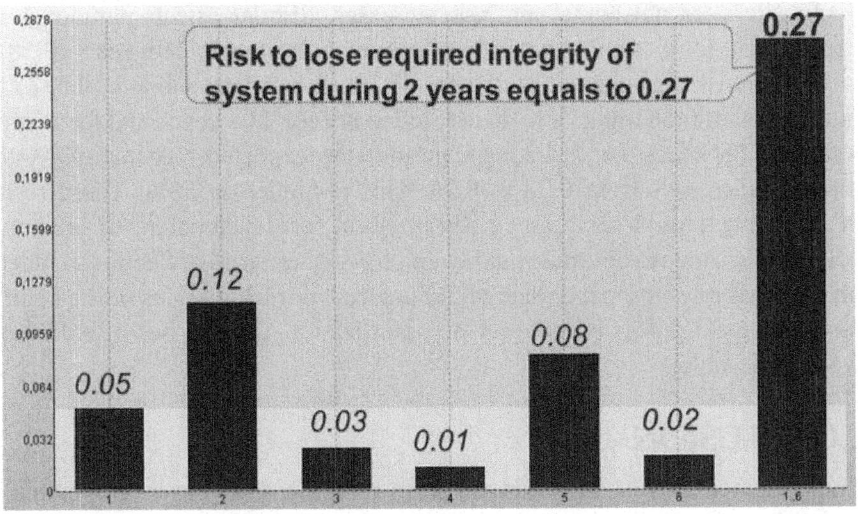

Figure 32: Risk to lose integrity for multipurpose system.

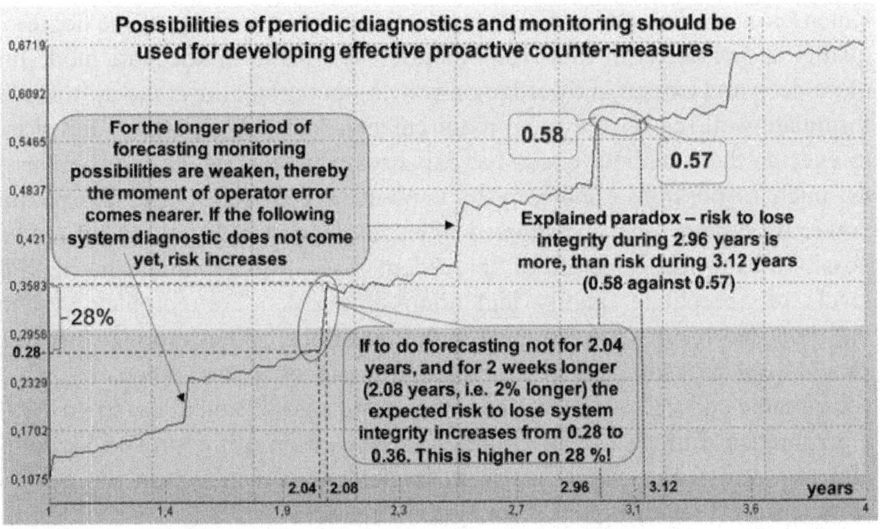

Figure 33: Integrated risk to lose integrity of system within operational 1 - 4 years.

Nonetheless, there is a lack of protection accumulated for the previous full periods that's why the risk doesn't decrease to 0 for every element. By the middle of a period between neighboring diagnostics there is an increase of the calculated risk because new danger sources can begin to influence. Moreover, for the longer period of prediction monitoring possibilities are weaken, thereby the moment of operator error comes nearer. And, if on timeline the following diagnostic does not come yet, risk increases. Similar effects paradoxes are explained— for example, that risk to lose integrity during 2.96 years (0.58) is more, than risk during more long time—3.12 years, 58 days longer (0.57). One more effect of modelling: if to do prediction not for 2.04 years, and for 2 weeks longer (2.08 years, i.e. 2% longer period) the expected risk to lose system integrity increases from 0.28 to 0.36. This is higher on 28%! These results of modeling should serve as a substantiation for development of predicting counter-measures to optimize quality and risks by criterias of Section 4. Indeed, on the basis of a rational choice of parametres for technologies of the control, monitoring and integrity recovery an optimization of processes offered in the work is possible.

CONCLUSION

The presented models, methods and software tools, allowing to predict quality and risks according to system requirements of standards, are real levers to analyze processes in system life cycle. The criteria for optimization are maximization of a prize (profit, a degree of quality or safety, etc.) at limits on expenses or minimization of expenses at limits on a comprehensible degree of quality and-or safety or their combination. As a result of adequate modelling more deep and extend knowledge of system operation allows the customer to formulate well-reasoned system requirements. And it is rational to developer to execute them without excessive expenses of resources, and to the user—as much as possible effectively to implement in practice the incorporated power of system. The investigated practical examples demonstrated models possibilities to use principle of "precedent cases" for definition the justified levels of acceptable quality and admissible risks. For complex systems the proposed results help to answer the questions "What rational measures should lead to estimated effect without waste expenses, when, by which controllable and uncontrollable conditions and costs?" and allow to go "from a pragmatical filtration of information to generation of the proved ideas and effective decisions". The effect from implementation in system life cycle is commensurable with expenses for system creation.

REFERENCES

1. A. I. Kostogryzov, A. V. Petuhov and A. M. Scherbina, "Foundations of Evaluation, Providing and Increasing Output Information Quality for Automatized System," Voorushenie. Politica. Konversija, Moscow, 1994.
2. A. I. Kostogryzov, "Software Tools Complex for Evaluation of Information Systems Operation Quality (CEISOQ)," Proceedings of the 34th Annual Event of the Government Electronics and Information Association

(GEIA), Engineering and Technical Management Symposium, Dallas, 2000, pp.63-70.

3. M. M. Bezkorovainy, A. I. Kostogryzov and V. M. Lvov, "Modelling Software Complex for Evaluation of Information Systems Operation Quality CEISOQ. 150 Problems of Analysis and Synthesis and Examples for Their Solutions," 2nd Edition, Voorushenie. Politica. Konversija, Moscow, 2002.
4. A. I. Kostogryzov and G. A. Nistratov, "Standardization, Mathematical Modelling, Rational Management and Certification in the Field of System and Software Engineering (80 Standards, 100 Mathematical Models, 35 Software Tools, More than 50 Practical Examples)," 2nd Edition, Voorushenie. Politica. Konversija, Moscow, 2005
5. A. I. Kostogryzov and G. A. Nistratov "100 Mathematical Models of System Processes According International Standards Requirements," Transaction of the XXV International Seminar on Stability Problems for the Stochastic Models, Maiority, 20-24 September 2005, pp. 196-201
6. A. Kostogryzov, G. Nistratov and N. Kleshchev, "Mathematical Models and Software Tools to Support an Assessment of Standard System Processes," Proceedings of the 6th International SPICE Conference on Process Assessment and Improvement (SPICE-2006), Luxembourg, 2006, pp. 63-68
7. A. Kostogryzov and G. Nistratov, "Mathematical Models and Software Tools for Analyzing System Quality and Risks According to Standard Requirements," Proceedings of the 6th International Scientific School, Modelling and Analysis of Safety and Risk in Complex Systems (MASR), Saint-Petersburg, RUSSIA, 4-8 July, 2006, pp. 155-163.
8. A. I. Kostogryzov and P. V. Stepanov, "Innovative Management of Quality and Risks in Systems Life Cycle," Voorushenie. Politica. Konversija, Moscow, 2008.
9. L. I. Grigoriev, V. Ya. Kershenbaum and A. I. Kostogryzov, "System Foundations of the Management of Competitiveness in Oil And Gas Complex," National Institute of Oil and Gas, Moscow, 2010.
10. A. Kostogryzov, V. Krylov, A. Nistratov, G. Nistratov, V. Popov and P. Stepanov, "Mathematical Models and Applicable Technologies to Predict, Analyze and Optimize Quality and Risks for Complex Systems," Proceedings of the 1st International Conference on Transportation Information and Safety (ICTIS 2011), 30 June-2 July 2011, Wuhan, pp. 845-854.

11. A. Kostogryzov, G. Nistratov and A. Nistratov, "Some Applicable Methods to Analyze and Optimize System Processes in Quality Management," In: T. Aized, Ed., Total Quality Management and Six Sigma, InTech, 2012, pp. 127-196, http://www.intechopen.com/books/total-quality-management-and-six-sigma/some-applicable-methods-to-analyze-and-optimize-system-processes-in-quality-management

12. A. Kostogryzov G. Nistratov and A. Nistratov, "Applicable Technologies to Predict, Analyze and Optimize Reliability and Risks for Complex Systems," Journal of Polish Safety and Reliability Association, SSARS, Vol. 3, No. 1, 2012, pp. 1-14.

13. W. Feller, "An Introduction to Probability Theory and Its Applications," Wiley, Hoboken, 1971.

14. B. V. Gnedenko, et al., "Priority Queueing Systems," Moscow State University, Moscow, 1973.

15. G. P. Klimov, "Probability Theory and Mathematical Statistics," Moscow State University, Moscow, 1983.

16. J. Martin, "System Analysis for Data Transmission," IBM System Research Institute, Prentice Hall, Inc., Englewood Cliffs, 1972.

17. L. Kleinrock, "Queueing Systems," John Wiley & Sons, New York, 1976.

18. V. F. Matweev and V. G. Ushakov, "Queuing Systems", Moscow State University, Moscow, 1984.

19. L. I. Machikhina, L. V. Alexeeva and L. S. L'vova, "Scientific Foundations of Food Grain Safety during Storage and Processing," DeLi Print, Moscow, 2007.

NOTES

[1]Note. System integrity is defined as such system state when system purposes are achieved with the required quality.

[2]Note. It is supposed that used diagnostic tools allow to provide necessary system integrity recovery after revealing of danger sources penetration into a system or consequences of influences.

Chapter 14

IMRT OPTIMIZATION WITH BOTH FRACTIONATION AND CUMULATIVE CONSTRAINTS

Delal Dink[1], Mark Langer[2], Seza Orcun[3], Joseph Pekny[1], Ronald Rardin[4], Gintaras Reklaitis[1], Behlul Saka[4]

[1]School of Chemical Engineering, Purdue University, W. Lafayette, USA

[2]Department of Radiation Oncology, Indiana University, Indianapolis, USA

[3]e-Enterprise Center, Discovery Park, Purdue University, W. Lafayette, USA

[4]Department of Industrial Engineering, University of Arkansas, Fayetteville, USA

ABSTRACT

Radiation therapy plans are optimized as a single treatment plan, but delivered over 30 - 50 treatment sessions (known as fractions). This paper proposes a new mixed-integer linear programming model to simultaneously incorporate fractionation and cumulative constraints in Intensity Modulated Radiation Therapy (IMRT) planning optimization used in cancer treatment. The method is compared against a standard practice of posing only cumulative limits in the optimization. In a prostate case, incorporating both forms of limits into planning converted an undeliverable plan obtained by considering only the cumulative limits into a deliverable one within 3% of the value obtained by ignoring the fraction size limits. A two-phase boosting strategy is studied as well, where the first phase aims to radiate primary and secondary targets simultaneously, and the second phase aims to escalate the tumor dose. Using of the simultaneous strategy on both phases, the dose difference between the primary and secondary targets was enhanced, with better sparing of the rectum and bladder.

INTRODUCTION

The success of radiation treatment rests on satisfying dose limits within critical or healthy structures. These criteria are determined mostly after years of experience and increasingly by formal clinical trials. Strictly, the prescriptions should be stated both in terms of cumulative doses as well as the dose received per treatment session known as fraction [1,2]. Both cumulative dose and per-fraction dose limits appear in the historical reports and clinical trial protocols that form the experience upon which modern treatments are based. In practice, planning optimization is based on cumulative dose limits alone. Numerous methods have been proposed in the literature to generate radiation therapy plans. Of these methods, optimization models using mathematical programming formulations have been developed to determine the best beamlet intensities [3-11] and the best aperture intensities [12,13]. Further methods include randomized approaches, such as simulated annealing [14-17] and genetic algorithms [18-20], and non-linear gradient techniques [21-24].

This paper investigates a conventional approach of optimizing over only cumulative limits as the optimization constraints alone and then dividing the plan into integral number of fractions after the optimization in order to obtain the daily delivery plan. We refer to this approach as Cumulative First Method. Then we present a new Simultaneous Method approach including both cumulative and fraction size limits simultaneously in the optimization which guarantee a plan that can be equally fractionated into a uniform course of the treatment. Both of the approaches try to achieve the uniform fractionation plan, where the same doses are delivered in every fraction.

The ability of the proposed approach to generate a solution that simultaneously satisfies the cumulative dose constraints and the fraction size constraints is tested via a test case, and the effect of not including the fraction sizes on the obtained objective, the minimum tumor dose and the normal structures is evaluated. By imposing both the cumulative and the fraction size limits into the formulation explicitly, the Simultaneous Method guarantees a solution when any feasible one exists by searching the feasible solution space with the fractionation criteria included. In addition to the two uniform fractionation methods, we also consider a two-phase approach familiar in conventional practice which uses different uniform fractions in two successive periods of treatment.

The primary and secondary targets are covered as uniformly as possible during the first phase of the treatment using the fraction sizes traditionally accepted for disease sterilization. Then, a second phase is added with the main focus to escalate the primary target. Relaxing the secondary target minimum fraction size requirements partially in the second phase of the treatment provides two potential advantages to the two-phase planning approach over the uniform fractionation: escalating the tumor dose or dropping the normal tissue exposure, if not both. Here, these potential advantages will be explored in Intensity Modulated Radiation Therapy (IMRT) planning where highly nonuniform beam patterns are possible by dividing the beam faces into small beamlets or bixels.

There have been studies where a non-IMRT module is accompanied by IMRT in the second phase for tumor boosting [25-27]. Two-phase IMRT planning was also tested where the coverage volume is changed from whole pelvic irradiation to prostate only in the second phase and it was concluded that when the whole pelvic area is simultaneously irradiated with IMRT higher dose to primary target and lower dose to the subclinical disease can be achieved [28]. Simultaneous optimization of the twophases for IMRT was addressed in [29] by using a gradient search algorithm with cumulative dose constraints alone. Here, we show the advantages of adapting our Simultaneous Method to two-phase planning.

A two-phase IMRT plan by irradiating the whole pelvic area is obtained with one-step optimization, where the irradiated volume is kept the same but the dose constraints for the two-phases are different and the dose difference between the primary disease and the secondary target is enhanced. The cumulative minimum dose limits on the secondary targets are satisfied during the first phase of the treatment. The goal of the second phase is solely to boost the tumor dose given the irradiated volume remains as the whole pelvic. In the second phase, it is no longer the interest of the planner to satisfy the minimum fraction size requirements of the secondary target since the potential disease on those sites are already eliminated during the first phase. Of course, the fraction sizes on the organs at risk remain to be a concern in the second phase.

This paper is organized as follows. Section 2 develops all the methods including Cumulative First Method, Simultaneous Method, and Two-Phase Planning with Simultaneous Method used to generate treatment plans in this paper. The results from computed plans are presented in Section 3 on a prostate case. Finally, the conclusions and possible future research directions are given in Section 4.

METHODS AND MATERIALS

In the formulations, each structure is identified as a primary target, a secondary target, or an organ at risk. The set of normal tissue indices is denoted by K, and the set of indices for the secondary target is denoted by V. The set of tissue points drawn from the primary target is denoted by T, that drawn from the k-th secondary target is denoted as S_k, and that drawn from the k-th organ at risk is denoted as H_k.

The solution approach in all cases employed a beamlet-based optimization in which dose to any point could be expressed as a linear combination of the individual beamlet intensities. Thus, dose d_i is the dose received at tissue point i in the system is defined as

$$\sum_{g \in G} \sum_{j \in J^g} a_{ijg} x_{jg} = d_i$$

$$\forall i \in T \bigcup \{S_k : k \in V\} \bigcup \{H_k : k \in K\} \qquad (1)$$

where x_{jg} is the value of total intensity assigned to beam let j of angle g, and a_{ijg} are pre-computed dose coefficients at point i from beam let j of beam g.

The chosen objective function in all approaches is to maximize the average tumor dose, a metric which has been found to be a predictor of tumor persistence after radiotherapy. It is represented as in (2) where |T| represents the total number of tumor points.

$$\text{Maximize} \ \sum_{i \in T} \frac{d_i}{|T|} \qquad (2)$$

In addition, preliminary work showed that the average tumor dose to be a computationally efficient surrogate for the minimum tumor dose in the presence of a strong tumor-dose homogeneity constraint. Both the value of the objective function and the value of the minimum tumor dose are evaluated in the results.

A homogeneity limit, defined as the ratio of the minimum to the maximum dose received, is enforced in all models to sustain a near uniform tumor dose profile.

$$d_{\min} \leq d_i \leq \frac{d_{\min}}{\alpha} \quad \forall i \in T \qquad (3)$$

Here $d_{\min}$ represents the minimum tumor dose value and parameter a > 0 is the allowed ratio. An alternative way of enforcing homogeneity in the tumor instead of specifying a ratio parameter α between the minimum and maximum tumor doses is to impose strict lower and upper bounds l_{tumor} and u_{tumor}, and replace relations (3) with (4) and (5).

$$d_i \geq l_{tumor} \quad \forall i \in T \qquad (4)$$

$$d_i \leq u_{tumor} \quad \forall i \in T \qquad (5)$$

Cumulative First Method

This conventional approach is based on the assumption that, if one can find a good plan to satisfy the cumulative dose limits, an appropriate number of fractions to indirectly satisfy the fraction size limits and to deliver the plan exists. Constraint sets (6) and (7) ensure that the maximum cumulative dose constraints for all healthy tissue points and minimum cumulative dose constraints for all secondary target points are satisfied. The superscript "total" represents the cumulative dose limits in the relations below. The limits l_{tumor} and u_{tumor} in the constraint sets (4) and (5) continue to represent the cumulative lower and upper bounds for the tumor.

$$d_i \le u_{ik}^{total} \quad \forall i \in H_k, k \in K \tag{6}$$

$$d_i \ge l_k^{total} \quad \forall i \in S_k, k \in V \tag{7}$$

Sometimes, dose volume limits are specified to limit the fractional portion p of a structure volume that can exceed a threshold dose in a normal tissue. A layering heuristic employed here to incorporate these limitations in the above model is as follows: After ordering the optimization points of the tissues with partial volume limits according to their distance to the tumor margin in 3D, the upper dose limits are assigned to that fraction p of the points that lie closest to the target boundary, and the lower dose limits are assigned to (1 – p) fraction of those points farthest from the target boundary [30].

Once a plan is developed based only on cumulative dose limits, it still must be divided into a fixed number of fractions which in today's practice are taken to be uniform. Practical bounds on N can be developed for the maximum and minimum number of allowed uniform fractions. An upper bound $\overline{N}$ comes from the minimum total doses achieved by tumor and secondary targets and dividing these numbers by the minimum allowed daily dose that should be delivered to each. The minimum of these ratios results in the upper bound for number of fractions given below in (8). Similarly, the lowest number $\underline{N}$ of allowed fractions is governed by the need not to exceed the fraction size requirement in any normal tissue (relation (9)).

$$N \le \overline{N} = \min_{i \in S_k, k \in V} \left\{ \frac{d_i}{l_k^{daily}}, \frac{d_{\min}}{l_{tumor}^{daily}} \right\} \tag{8}$$

$$N \ge \underline{N} = \max_{i \in H_k, k \in K} \left\{ \frac{d_i}{u_k^{daily}} \right\} \tag{9}$$

The number of fractions N for a plan should satisfy $\underline{N} \leq N \leq \overline{N}$. If these fraction size requirements cannot be satisfied for any integral N, then the plan produced without consideration of the fraction size requirements is infeasible. Furthermore, a plan for which no integer number of fractions will satisfy bounds (8) and (9) on the fraction size cannot be renormalized to satisfy the fraction size requirements. Violation $\underline{N} > N$ means from (8) and (9) that

$$\max_{i\in H_k, k\in K}\left\{\frac{d_i}{u_k^{daily}}\right\} > \min_{i\in S_k, k\in V}\left\{\frac{d_i}{l_k^{daily}}, \frac{d_{\min}}{l_{tumor}^{daily}}\right\} \tag{10}$$

Thus, renormalizing the plan by scaling down the beam let intensities, x_{jg}, and thus total doses i d by factor l > 0, has only the effect of multiplying both sides of (10) by the same l > 0. The infeasibility would remain.

Simultaneous Method

Our proposed Simultaneous Method, unlike the earlier approach, optimizes over both the cumulative and per fraction dose limits at the same time, where the total number of fractions are included explicitly [31]. The complete problem was specified as a linear or mixed integer program, in which doses in each tissue must fall within an allowed range over each treatment session, and the cumulative dose distribution across each structure must meet the specified dose, dose-volume and homogeneity limits.

In order to achieve this, objective function (2), homogeneity limits (3) or (4)-(5) combined with cumulative constraints (6) and (7) are retained, and new constraints (11)-(13) are introduced as follows:

$$d_i \leq N \times u_k^{daily} \quad \forall i \in H_k, k \in K \tag{11}$$

$$d_i \geq N \times l_k^{daily} \quad \forall i \in S_k, k \in V \tag{12}$$

$$d_i \geq N \times l_{tumor}^{daily} \quad \forall i \in T \tag{13}$$

Here, N represents the total integer number of treatment sessions. Upper and the lower fraction doses and limits for tumor, secondary targets, and healthy tissues along with per-fraction doses are denoted by the superscript "daily" assuming one fraction is delivered per day.

The effect of constraints (11)-(13) is to ensure that total doses i accommodate N times the applicable per-fraction limit on each tissue. Then a feasible fraction plan can be obtained simply by dividing all doses (equivalently intensities) by

N. We choose to treat N as an integer variable and optimize the number of fractions.

Two-Phase Planning with Simultaneous Method

In the two-phase planning approach, there are different plans and numbers of fractions for the two phases. We extend our Simultaneous Method by adding subscripts $e = 1, 2$ to distinguish doses, intensities, and the numbers of fractions for the first and second periods respectively. Table 1 details the mixed-integer program.

In Table 1, constraint (15) puts an upper bound on cumulative healthy tissue dose, (16) enforces tumor homogeneity, (17) and (18) limit the maximum healthy tissue dose per each fraction of the two phases, and (19)- (20) set maximum tumor fraction size in each phase. Constraints (21) and (22) applies cumulative and per-fraction constraints on secondary targets only in phase 1. It should be noted that the only integer variables of the above method are N1 and N2, which makes the mixed-integer linear programming (MILP) formulation a relatively easy one to handle. The method is tested with the same prostate case and compared to uniform fractionation scenario.

CASE STUDIES

CPLEX with branch and bound algorithm was employed for the following test case. The points for optimization were distributed throughout the contours, determined randomly within a structure volume for computation efficiency for optimization rather than employing a uniform point set, yet highly concentrated especially within the target and the critical structures of interest. [32-36]. For each structure, we presented numbers of points used in sampling and computed the mean distances to the nearest neighbor point. Further details are provided below. The influence matrix was calculated by using the standard radiation therapy software GRATIS [37].

Effect of Including Fraction Size Post-Optimally with Cumulative First

The effect of Cumulative First Method, i.e. not including the fraction size constraints in optimization is examined for a case of carcinoma of prostate. The data set consisted of 23 CT slices of which 5 are illustrated in Figure 1. The prostate case contained a primary target volume identified as the prostate gland, a secondary target volume consisting of the seminal vesicles, and four healthy tissue structures, namely the bladder, two femoral heads, rectum and all other surrounding unspecified normal tissue. Number of sample points used (the mean distance to the nearest neighbor point) is 400 (0.44 cm) for bladder, 1000 (0.38 cm) for both femoral heads together, 1000 (0.28 cm) for rectum, 465 (1.4 cm) for the remaining unspecified tissue, 786 (0.29 cm) for PTV1 (Planning Target Volume 1), and 252 (0.24 cm) for the seminal vesicles. The prescriptions for both cumulative and fraction doses are given in Table 2.

Table 1: Two-phase planning with simultaneous model

Cumulative Conditions			
Maximize $\sum_{e\in\{1,2\}}\sum_{i\in T} d_{ie}$			(14)
$\sum_{e\in\{1,2\}} d_{ie} \le u_{ik}^{total} \quad \forall i \in H_k, k \in K$			(15)
$d_{\min} \le \sum_{e\in\{1,2\}} d_{ie} \le \frac{d_{\min}}{\alpha} \quad \forall i \in T$			(16)
Phase 1 Conditions		**Phase 2 Conditions**	
$d_{i1} \le u_k^{daily} \times N_1 \quad \forall i \in H_k, k \in K$	(17)	$d_{i2} \le u_k^{daily} \times N_2 \quad \forall i \in H_k, k \in K$	(18)
$d_{i1} \ge l_{tumor}^{daily} \times N_1 \quad \forall i \in T$	(19)	$d_{i2} \ge l_{tumor}^{daily} \times N_2 \quad \forall i \in T$	(20)
$d_{i1} \ge l_k^{total} \quad \forall i \in S_k, k \in V$	(21)		
$d_{i1} \ge l_k^{daily} \quad \forall i \in S_k, k \in V$	(22)		
N_1, N_2 nonnegative integers			(23)

Table 2: Prostate case study prescription

Site	Volume Definition	Cumulative Dose Objective (Gy)	Dose/fx Objective (Gy)	Restricted Volume
Prostate Tumor	PTV1	≤84 ≥79	≥2	100%
Seminal Vesicles	PTV2	≤84 ≥56	≥2	100%
Bladder	Normal	≤85 ≤80	≤2	100% 80%
Femoral Heads (2)	Normal	≤72 ≤50	≤2	100% 60%
Rectum	Normal	≤85 ≤80	≤2	100% 80%
Other unspecified normal tissue	Normal	≤85	≤2	100%

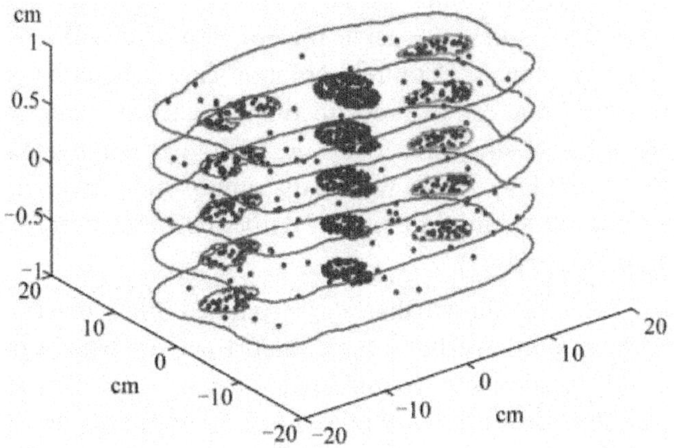

Figure 1: Outlines of 5 of the 23 transverse slices of the prostate case taken through the inferior portion of the treatment volume, with optimization sample points shown within structures of interest. The densest collection of points is taken within and on the boundary of the prostate target. Also shown are samplings from the two femurs, the rectum, and the unspecified normal tissue within the external surface.

The prescription includes a set of limits on the maximum doses that can be received in a structure as well as partial volume limits placed on the bladder, rectum and femoral heads. Those constraints are reduced to pointby-point upper limits on those tissues by protecting the points farthest from the tumor as explained in Subsection 2.1. Table 2 also includes upper or lower bounds on the doses that can be given to structures of interest each fraction. A lower bound on the dose per fraction to PTV1 and PTV2 (Planning Target Volumes 1 & 2), and an upper bound on the dose per fraction to any normal tissue structure are set at 2 Gy. Tumor dose homogeneity is required as in constraints (4) and (5) by a minimum dose limit of 79 Gy and maximum dose limit of 84 Gy to PTV1. In addition to the minimum cumulative dose limit on secondary target, a maximum dose limit of 84 Gy is also assigned to ensure that the surrounding structures of the tumor remain below the maximum tumor dose. The number of fields (beam angels) used in the plan is 9, with the angles pre-specified.

A treatment plan was prepared first with Cumulative First Method without regarding to the fraction size limits, but only regarding to the cumulative dose limits, using the optimization model described in Subsection 2.1. The resulting dose-volume histograms (DVH) for the target, bladder and rectum are shown in Figure 2. The DVHs show the dose received by the volume percentage of the critical structures. All of the extreme dose and partial dose volume conditions expressed in Table 2 were satisfied over the points employed in the optimization routine.

Table 3 shows the values obtained for the extreme doses delivered to target or normal tissues alongside the fraction size and cumulative dose limits which apply to them. The final column shows the number of fractions (treatment sessions) that can be prescribed under each separate condition for a normal tissue or target structure, found by applying relations (8) and (9), and substituting D05 and D95 values for the maximum and minimum doses. Here, D05 and D95 denote the minimum dose delivered to the hottest 5% and 95% of the structure, respectively.

Scanning Table 3 shows that the produced plan meets the required cumulative dose conditions, but a number of fractions cannot be determined such that all the fraction size requirements are met. The unspecified tissue, bladder and rectum conditions require more fractions to be delivered than the maximum allowed by the condition on the secondary target (PTV2), and bladder and rectum require more fractions than allowed by the primary target. For example, the bladder requires a minimum of 42 fractions to satisfy its fraction size goals with this plan, while the PTV2 requires a maximum of 28 fractions. Renormalizing this plan to satisfy the fraction size requirements is not possible for any number of treatment sessions as explained above.

Instead of trying to fractionate the plan after the optimization, the fractionation was imposed as part of the optimization problem for the same case using the Simultaneous Method of Subsection 2.2. The results are given in the form of dose-volume histograms for target, bladder and rectum in Figure 3. The total number of fractions, N, is optimized and found that the plan obtained with both fraction size and cumulative dose limits can be delivered in 41 equal fractions.

Table 3: Dose statistics of the prostate case obtained by optimizing with cumulative first method (Disregarding fraction constraints)

Structure	Dose Statistics	Required Cumulative Dose (Gy)	Achieved Cumulative Dose (Gy)	Dose per Fraction Limit (Gy)	Resulting Bound on Number of Fractions
Target	Min (D95)	≤84.0 ≥79.0	83.88	≥2.0	≤41
PTV2	Min (D95)	≤84.0 ≥56.0	56.09	≥2.0	≤28
Bladder	Max (D05)	≤85.0	83.74	≤2.0	≥42
FemHead1	Max (D05)	≤72.0	48.50	≤2.0	≥25
FemHead2	Max (D05)	≤72.0	47.12	≤2.0	≥24
Rectum	Max (D05)	≤85.0	82.99	≤2.0	≥42
Otherwise unspecified tissue	Max (D05)	≤85.0	63.74	≤2.0	≥32

Table 4: Comparison of cumulative dose statistics of the prostate case given by cumulative first and simultaneous methods

Structure	Statistics	Cumulative First Method (Gy)	Simultaneous Method (Gy)
	Max Dose (D05)	84.00	84.00
Target	Min Dose (D95)	83.88	82.91
	Mean Dose	83.97	83.75
PTV2	Min Dose (D95)	56.09	82.00
Bladder	Max Dose (D05)	83.74	81.97
FemHead1	Max Dose (D05)	48.50	48.66
FemHead2	Max Dose (D05)	47.12	50.68
Rectum	Max Dose (D05)	82.99	80.74
Otherwise unspecified tissue	Max Dose (D05)	63.74	63.06

Table 4 compares the relevant statistics on the extreme doses received in targets and healthy tissues under with and without fraction size limits. With either model, both the extreme dose values and the partial volume doses satisfied their required limits. As can be seen, the minimum primary target dose in the model that satisfied the fraction size limits was only 1.2% less (83.88 vs 82.91 Gy) than the infeasible solution found by removing the fraction size requirement.

Comparing Figure 2 and Figure 3, there are no significant differences in the rectum and target dose distribution; however the bladder dose in the 60% - 20% volume region increases when fraction size limits are included to the planning. Still, when the extreme doses are considered, the drop in the bladder and rectal doses are greater than the 0.26% drop in the primary target dose, drops of 1.77 Gy and 2.25 Gy respectively.

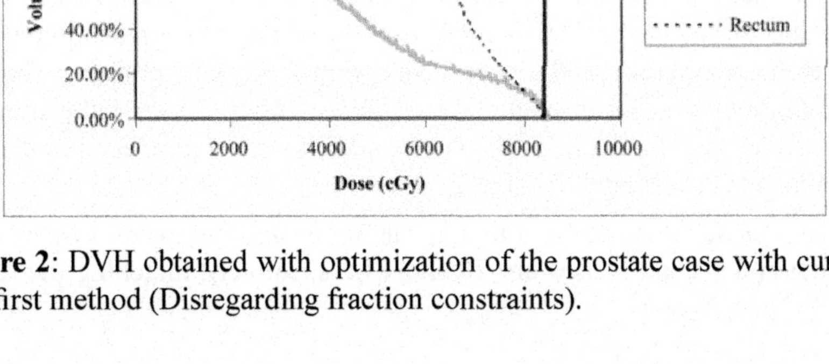

Figure 2: DVH obtained with optimization of the prostate case with cumulative first method (Disregarding fraction constraints).

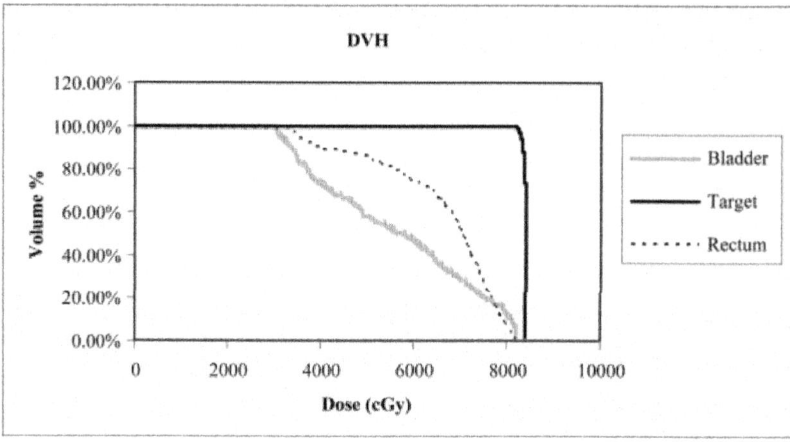

Figure 3: DVH obtained with optimization of the prostate case with fraction size and cumulative dose limits together (Simultaneous method).

Noteworthy is the change in the dose delivered to the secondary target (PTV2). The lower dose bound for each treatment session of 2.0 Gy placed on this structure could not be satisfied simultaneously with an upper dose bound for each session of 2.0 Gy placed on normal tissues if the model that eliminated the fractionation terms were used. Once the fractionation terms were added, the minimum dose in the PTV1 fell by 0.26% but the minimum dose in the PTV2 rose from 56 Gy to 82 Gy.

A comparison of isodose plots for the plans constructed with and without a limit on fraction sizes delivered to the critical tissues is shown in Figures 4(a) and (b). The isodose plots are very similar, showing that a deliverable plan can be obtained reaching the same cumulative doses when the fractionation considerations are included in the optimization.

Two-Phase Planning versus Uniform Fractionation with Simultaneous Method

The prostate case studied above with uniform fractionation with simultaneous method is further analyzed with the two-phase formulation given in Subsection 2.3. Both cumulative and fraction size limits are still enforced, however the minimum fraction size limits on PTV2 is removed after the minimum cumulative dose limit of 56 Gy is reached in each and every point of this structure as in the model in Table 1.

The results are given in Table 5. The optimized numbers of fractions for the first and second phases are 28 and 13 respectively, matching the total of

41 equal fractions with the previous plan. Keeping in mind that there is a minimum and maximum dose limit on the primary target PTV1, insignificant difference is observed on the PTV1 dose distribution between the two methods. However, the decrease in the PTV2 seminal vesicles is quite noteworthy. It can be observed further in detail in the dose-volume histograms given in Figure 5 that delivering the plan in two different phases helps to spare the vesicles from significant amount of radiation.

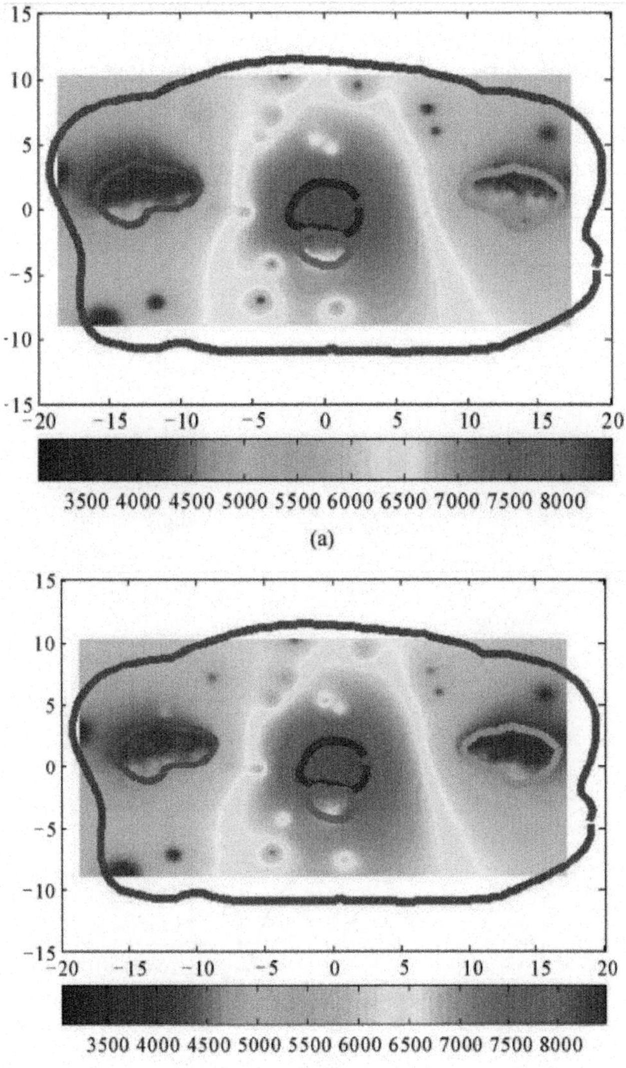

Figure 4: (a) Isodose surfaces for the prostate plan obtained with cumulative first method; (b) Isodose surfaces for the prostate plan obtained with simultaneous method.

Unnecessary exposure to radiation in those irradiated areas increases the potential of complications at the end of the treatment, although there is a minimum cumulative limit on the secondary targets; it is not the intention of the planner to radiate them above 56 Gy. Furthermore, although the extreme values do not show much difference between the two approaches, the dose volume histograms given in Figure 6 demonstrate that removing fraction size requirement from seminal vesicles during the second phase of the treatment helps to reduce the bladder dose on average significantly and rectum exposure slightly as well. Although the extreme dose limits on these two organs at risk are almost the same with the two methods, the doses at different volume percentages are quite far apart.

Table 5: Comparative results of the prostate case with the two-phase model

Structure	Statistics	Uniform Fractionation with Simultaneous Method (Gy)	Two-Phase Model (Gy)
	Max Dose (D05)	84.00	84.00
Target	Min Dose (D95)	82.91	82.92
	Mean Dose	83.75	83.76
	Max Dose (D05)	84.00	83.46
PTV2	Min Dose (D95)	82.00	64.90
	Mean Dose	82.98	75.70
Bladder	Max Dose (D05)	81.97	81.73
FemHead1	Max Dose (D05)	48.66	47.38
FemHead2	Max Dose (D05)	50.68	49.09
Rectum	Max Dose (D05)	80.74	80.37
Otherwise unspecified tissue	Max Dose (D05)	63.06	64.59
Number of fractions		41	28 + 13

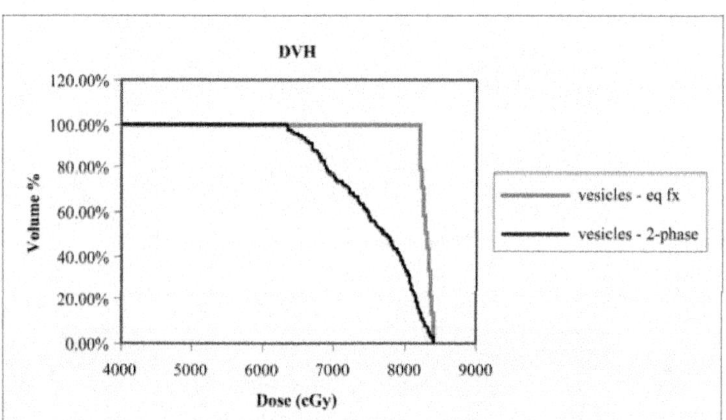

Figure 5: Comparison of DVH of vesicles for uniform fractionation and two-phase simultaneous models.

The isodose surfaces given in Figure 7 for the two approaches provide another tool to observe the improvement with the two-phase model. The plots on the left correspond to the uniformly fractionated plan with the simultaneous method, while the ones on the right are obtained with two phase planning. The color scales on all four plots are kept the same for fair comparison. Although there is no apparent difference on the transverse slice at z = 0, the second slice at z = 4.4 which includes the bladder, rectum and the seminal vesicles demonstrate the decrease in radiation exposure in all parts of the irradiated volume.

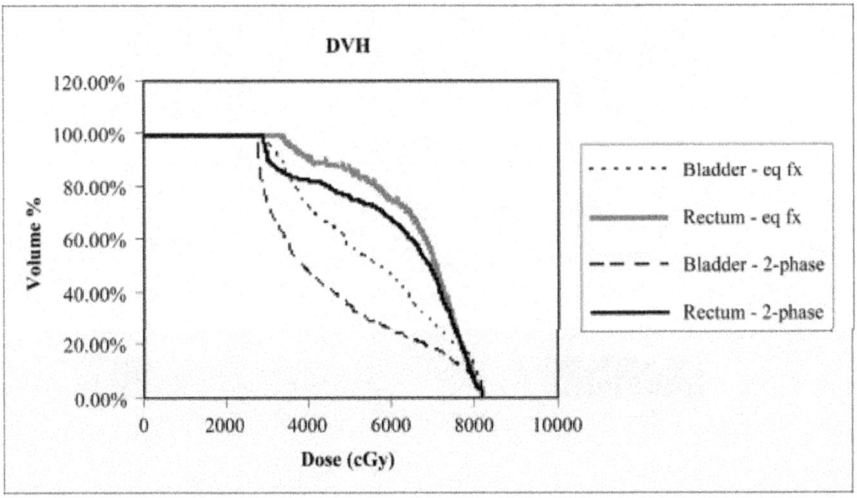

Figure 6: Comparison of DVH of bladder and rectum for uniform fractionation and two-phase simultaneous models.

DISCUSSIONS AND CONCLUSIONS

In this paper, the problem of producing an IMRT plan that meets fraction size and cumulative dose limits on tumor and normal tissues simultaneously is addressed. The common approach of employing the cumulative dose limits alone and considering fractionation as a post-optimization stage, i.e. The Cumulative First approach, was demonstrated on a prostate case for which the dose distribution planned without considering fraction size requirements could not subsequently be partitioned into uniform sessions so as to satisfy the limits. It was observed that a plan that is feasible with respect to the accumulated doses may no longer remain feasible with uniform fractions once upper and lower fraction size limits are specified. The objective obtained upon imposing fraction size and cumulative limits was only 0.26% less than that which could be found even if no fraction size requirements were posed, a difference smaller than the uncertainty in the delivered dose or in the minimum dose that can be delivered in a treatment session. Hence, both cumulative and fraction size limits could be used in the planning and the fraction size requirements could be accommodated without materially losing the level of dose escalation which can be achieved with IMRT.

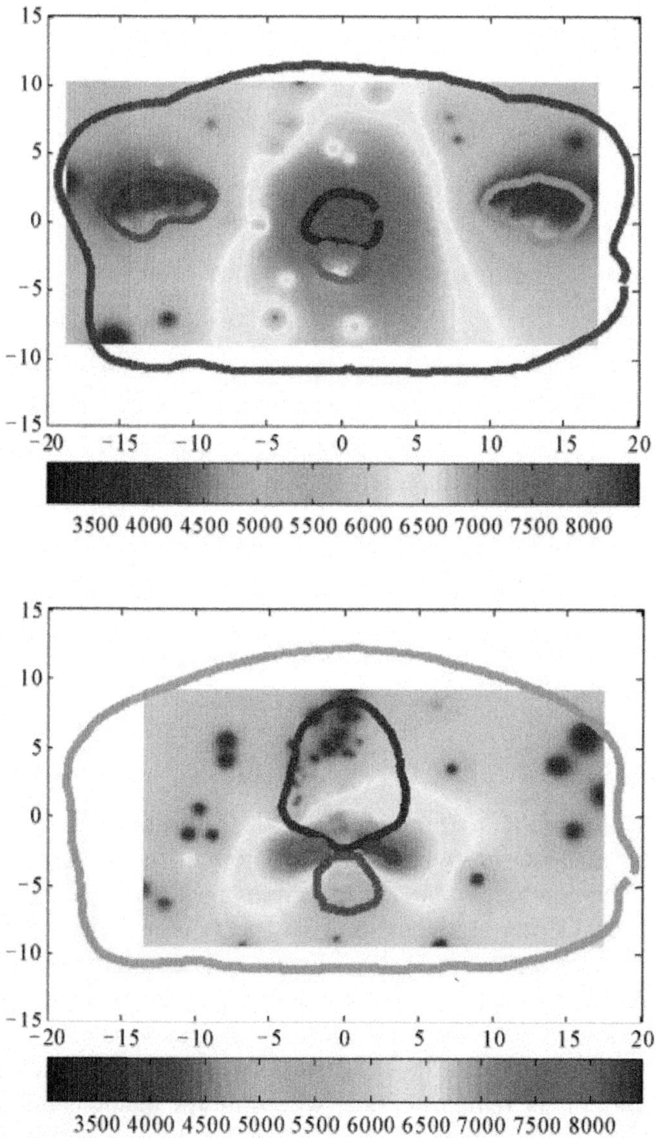

(a)

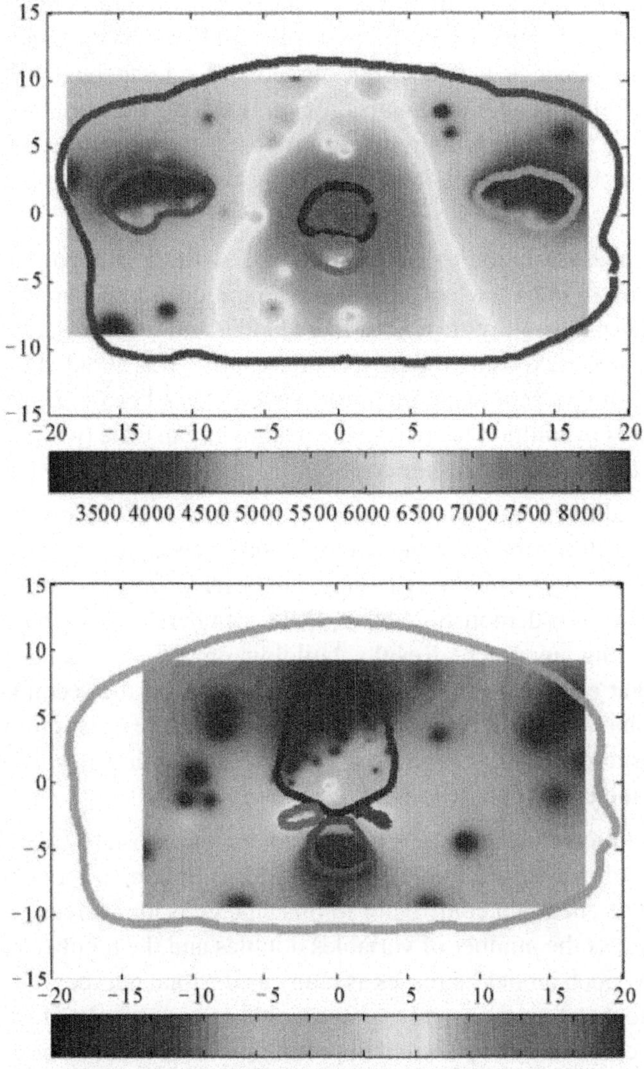

(b)

Figure 7: Isodose surfaces for transverse slices of z = 0 and z = 4.4 prostate case with (a) Simultaneous uniform fractionation (b) Simultaneous two-phase planning.

In operations research, it is common practice to represent a problem with the least constraints possible by leaving dependent equations out to reduce solution time. However, as the experience of this research showed, the cumulative dose and fraction size limits are neither dependent nor redundant, and none of them should be eliminated from the formulation.

Further, an analysis of bounds proved that no renormalization of the prescription would improve the plans developed based on the cumulative limits alone. In the paper, it was also questioned if Simultaneous IMRT approach would benefit from combining with a conventional planning method: two-phase planning. An MILP formulation was proposed incorporating all cumulative and fraction size dose limits and optimizing the number of fractions for each phase.

A two-phase solution to the prostate case without changing the irradiated margins was found in which the same tumor dose exposure was achieved while the dose to healthy structures like bladder and rectum were lessened and unwanted further exposure on the secondary target was avoided in comparison to the uniformly fractionated plan. Moreover, it has been our experience that in some cases in addition to sparing the healthy organs, the two-phase planning also helps to escalate the tumor dose.

One might assume that the detailed intensity modulation achieved with multi-leaf collimators or other modulation techniques might not require partitioning the problem into two phases in order to improve the quality of the treatment. Yet, it is demonstrated that, IMRT can benefit from the conventional tumor boosting approach. Results building on the simultaneous approach establish that releasing the fraction size minimum on the secondary target in a second phase of the therapy can favor tumor dose escalation or reduce the normal tissue exposure. Hence, the two-phase planning is still a method to explore in the era of detailed intensity maps.

The solution times are also compared for the prostate case. It increases from order of minutes to hours for the uniform fractionation versus the two-phase model. The main contributor to this change is the increase in the size of the problem, as the number of variables doubles and the number of constraints for the two epoch model increases as many as the total number of tissue points, which is close to four thousand in this specific case. It requires further research to employ efficient heuristics and approximations in order to benefit from the simultaneous and two-phase approaches in the clinics in practice. It should be noted that, although, the optimization method used here is linear programming, there is no reason not to think that the same conclusions could be reached when this analysis of simultaneous and two-phase methods was repeated with other optimization tools.

Finally, although the two-phase approach shows promise as it is, it would be of interest to the therapy planner to investigate the possibility of allowing different beam orientations during the two phases of the therapy.

ACKNOWLEDGEMENTS

We would like to thank Seda Gumrukcu for her contribution in managing submissions of this work.

REFERENCES

1. Q. Wu, M. Manning, R. Schmidt-Ulrich and R. Mohan, "The Potential for Sparing of Parotids and Escalation of Biologically Effective Dose with Intensity-Modulated Radiation Treatments of Head and Neck Cancers: A Treatment Design Study," International Journal of Radiation Oncology, Biology, Physics, Vol. 46, No. 1, 2000, pp. 195-205. doi:10.1016/S0360-3016(99)00304-1
2. A. I. Blanco and K. S. C. Chao, "Intensity-Modulated Radiation Therapy and Protection of Normal Tissue Function in Head and Neck Cancer," In: C. A. Perez and L. W. Brady, Eds., Principles and Practice of Radiation Oncology Updates, Lippincott Williams & Wilkins Healthcare, New York, 2002, pp. 2-12.
3. M. Langer, R. Brown, M. Urie, J. Leong, M. Stracher and J. Shapiro, "Large Scale Optimization of Beam Weights under Dose-Volume Restrictions," International Journal of Radiation Oncology, Biology, Physics, Vol. 18, No. 4, 1990, pp. 887-893. doi:10.1016/0360-3016(90)90413-E
4. M. Langer, P. Kijewski, R. Brown and C. Ha, "The Effect on Minimum Tumor Dose of Restricting Target-Dose Inhomogeneity in Optimized Three-Dimensional Treatment of Lung Cancer," Radiotherapy and Oncology, Vol. 21, No. 4, 1991, pp. 245-256. doi:10.1016/0167-8140(91)90049-M
5. M. Langer, E. K. Lee, J. O. Deasy, R. L. Rardin and J. A. Deye, "Operations Research Applied to Radiotherapy, An NCI-NSF-Sponsored Workshop," International Journal of Radiation Oncology, Biology, Physics, Vol. 57, No. 3, 2003, pp. 762-768. doi:10.1016/S0360-3016(03)00720-X
6. E. K. Lee, T. Fox and I. Crocker, "Integer Programming Applied to Intensity-Modulated Radiation Treatment Planning," Annals of Operations Research, Vol. 119, No. 1-4, 2003, pp. 165-181. doi:10.1023/A:1022938707934
7. E. K. Lee, T. Fox and I. Crocker, "Simultaneous Beam Geometry and Intensity Map Optimization in Intensity-Modulated Radiation Therapy," International Journal of Radiation Oncology, Biology, Physics, Vol. 64, No. 1, 2006, pp. 301-320. doi:10.1016/j.ijrobp.2005.08.023

8. H. E. Romeijn, R. K. Ahuja, J. F. Dempsey, A. Kumar and J. G. Li, "A Novel Linear Programming Approach to Fluence Map Optimization for Intensity Modulated Radiation Treatment Planning," Physics in Medicine and Biology, Vol. 48, No. 21, 2003, pp. 3521-3542. doi:10.1088/0031-9155/48/21/005

9. H. E. Romeijn, R. K. Ahuja and J. F. Dempsey, "A New Linear Programming Approach to Radiation Therapy Treatment Planning Problems," Operations Research, Vol. 54, No. 2, 2006, pp. 201-216. doi:10.1287/opre.1050.0261

10. F. Preciado-Walters, R. L. Rardin, M. P. Langer and V. Thai, "A Coupled Column Generation, Mixed Integer Approach to Optimal Planning of Intensity Modulated Radiation Therapy for Cancer," Mathematical Programming, Vol. 101, No. 2, 2004, pp. 319-338. doi:10.1007/s10107-004-0527-6

11. A. Tuncel, "Methods and Algorithms for Radiation Therapy Optimization under Dose-Volume Restrictions," Ph.D. Dissertation, Purdue University, West Lafayette, 2008.

12. F. Preciado-Walters, M. P. Langer, R. L. Rardin and V. Thai, "Column Generation for IMRT Cancer Therapy Optimization with Implementable Segments," Annals of Operations Research, Vol. 148, No. 1, 2006, pp. 65-79. doi:10.1007/s10479-006-0080-1

13. H. E. Romeijn, R. K. Ahuja, J. F. Dempsey and A. Kumar, "A Column Generation Approach to Radiation Therapy Treatment Planning Using Aperture Modulation," SIAM Journal on Optimization, Vol. 15, No. 3, 2005, pp. 838-862. doi:10.1137/040606612

14. S. Webb, "Optimization by Simulated Annealing of ThreeDimensional Conformal Treatment Planning for Radiation Fields Defined by a Multileaf Collimator," Physics in Medicine and Biology, Vol. 36, No. 9, 1991, pp. 1201- 1226. doi:10.1088/0031-9155/36/9/004

15. S. M. Morril, R. G. Lane and I. I. Rosen, "Constrained Simulated Annealing for Optimized Radiation Therapy," Computer Methods and Programs in Biomedicine, Vol. 33, No. 3, 1990, pp. 135-144. doi:10.1016/0169-2607(90)90035-8

16. G. S. Mageras and R. Mohan, "Application of Fast Simulated Annealing to Optimization of Conformal Ra diation Treatments," Medical Physics, Vol. 20, No. 3, 1993, pp. 639-647. doi:10.1118/1.597012

17. M. Langer, S. Morrill, R. Brown, O. Lee and R. Lane, "A Comparison of Mixed-Integer Programming and Fast Simulated Annealing for

Optimizing Beam Weights in Radiation Therapy," Medical Physics, Vol. 23, No. 6, 1996, pp. 957-964. doi:10.1118/1.597857

18. M. Langer, S. Morrill, R. Brown, O. Lee and R. Lane, "A Genetic Algorithm for Generating Beam Weights," Medical Physics, Vol. 23, No. 6, 1996, pp. 965-971. doi:10.1118/1.597858
19. G. A. Ezzel, "Genetic and Geometric Optimization of Three-Dimensional Radiation Therapy Treatment Planning," Medical Physics, Vol. 23, No. 3, 1996, pp. 293- 305. doi:10.1118/1.597660
20. X. Wu, Y. Zhu, J. Dai and Z. Wang, "Selection and Determination of Beam Weights based on Genetic Algorithms for Conformal Radiotherapy Treatment Planning," Physics in Medicine and Biology, Vol. 45, No. 9, 2000, pp. 2547-2558. doi:10.1088/0031-9155/45/9/308
21. P. S. Cho, S. Lee, R. J. Marks, S. Oh, S. G. Sutlief and M. H. Phillips, "Optimization of Intensity Modulated Beams with Volume Constraints Using Two Methods: Cost Function Minimization and Projections onto Convex Sets," Medical Physics, Vol. 25, No. 4, 1998, pp. 435- 443. doi:10.1118/1.598218
22. D. H. Hristov and B. G. Fallone, "A Continuous Penalty Function Method for Inverse Treatment Planning," Medical Physics, Vol. 25, No. 2, 1998, pp. 208-223. doi:10.1118/1.598183
23. S. V. Spirou and C. S. Chui, "A Gradient Inverse Planning Algorithm with Dose-Volume Constraints," Medical Physics, Vol. 25, No. 3, 1998, pp. 321-333. doi:10.1118/1.598202
24. Q. Wu and R. Mohan, "Algorithms and Functionality of an Intensity Modulated Radiotherapy Optimization System," Medical Physics, Vol. 27, No. 4, 2000, pp. 701- 711. doi:10.1118/1.598932
25. P. Cheung, K. Sixel, G. Morton, D. A. Loblaw, R. Tirona, G. Pang, R. Choo, E. Szumacher, G. Deboer and J. P. Pignol, "Individualized Planning Target Volumes for Intrafraction Motion during Hypofractionated IntensityModulated Radiotherapy Boost for Prostate Cancer," International Journal of Radiation Oncology, Biology, Physics, Vol. 62, No. 2, 2005, pp. 418-425.
26. M. Guerrero, X. A. Li, L. Ma, J. Linder, C. Deyoung and B. Erickson, "Simultaneous Integrated Intensity-Modulated Radiotherapy Boost for Locally Advanced Gynecological Cancer: Radiobiological and Dosimetric Considerations," International Journal of Radiation Oncology, Biology, Physics, Vol. 62, No. 3, 2005, pp. 933-939. doi:10.1016/j.ijrobp.2004.11.040

27. M. L. Cavey, J. E. Bayouth, M. Colman, E. J. Endres and G. Sanguineti, "IMRT to Escalate the Dose to the Prostate While Treating the Pelvic Nodes," Strahlenther Onkol, Vol. 181, No. 7, 2005, pp. 431-441. doi:10.1007/s00066-005-1384-9

28. X. A. Li, J. Z. Wang, P. A. Jursinic, C. A. Lawton and D. Wang, "Dosimetric Advantages of IMRT Simultaneous Integrated Boost for High-Risk Prostate Cancer," International Journal of Radiation Oncology, Biology, Physics, Vol. 61, No. 4, 2005, pp. 1251-1257. doi:10.1016/j.ijrobp.2004.11.034

29. R. A. Popple, P. B. Prellop, S. A. Spencer, J. F. De Los Santos, J. Duan, J. B. Fiveash and I. A. Brezovich, "Simultaneous Optimization of Sequential IMRT Plans," Medical Physics, Vol. 32, No. 11, 2005, pp. 3257-3266. doi:10.1118/1.2064849

30. S. M. Morrill, R. G. Lane, J. A. Wong and I. I. Rosen, "Dose-Volume Considerations with Linear Programming Optimization," Medical Physics, Vol. 18, No. 6, 1991, pp. 1201-1210. doi:10.1118/1.596592

31. D. Dink, "Approaches to 4-D Intensity Modulated Radiation Therapy Planning with Fraction Constraints," Ph.D. Dissertation, Purdue University, West Lafayette, 2005.

32. S. M. Morrill, I. I. Rosen, R. G. Lane and J. A. Belli, "The Influence of Dose Constraint Point Placement on Optimized Radiation Therapy Treatment Planning," International Journal of Radiation Oncology, Biology, Physics, Vol. 19, No. 1, 1990, pp. 129-1241.

33. A. Niemierko and M. Goitein, "Random Sampling for Evaluating Treatment Plans," Medical Physics, Vol. 17, No. 5, 1990, pp. 753-762. doi:10.1118/1.596473

34. A. Niemierko and M. Goitein, "Letter to the Editor, Comments on 'Sampling Techniques for the Evaluation of Treatment Plans'," Medical Physics, Vol. 20, No. 1, 1993, pp. 1377-1380. doi:10.1118/1.597103

35. X.-Q. Lu and L. M. Chin, "Sampling Techniques for the Evaluation of Treatment Plans," Medical Physics, Vol. 20, No. 1, 1993, pp. 151-161. doi:10.1118/1.597096

36. R. Acosta, W. Brick, A. Hanna, A. Holder, D. Lara, G. McQuilen, D. Nevin, P. Uhlig and B. Salter, "Radiotherapy Optimal Design: An Academic Radiotherapy Treatment Design System," In: J. W. Chinneck, B. Kristjansson, and M. J. Saltman, Eds., Operations Research and Cyber-Infrastructure, Springer, New York, 2009, pp. 401- 425. doi:10.1007/978-0-387-88843-9_21

37. Sherouse Systems Inc., 2011. http://www.gwsherouse.com

CITATION

CHAPTER 1

Z. Zhang, M. Liao and L. Wang, "Immune Optimization Approach for Dynamic Constrained Multi-Objective Multimodal Optimization Problems," *American Journal of Operations Research*, Vol. 2 No. 2, 2012, pp. 193-202. doi: 10.4236/ajor.2012.22022.

CHAPTER 2

M. Frutos and F. Tohmé, "A New Approach to the Optimization of the CVRP through Genetic Algorithms," American Journal of Operations Research, Vol. 2 No. 4, 2012, pp. 495-501. doi: 10.4236/ajor.2012.24058.

CHAPTER 3

S. Sim and H. Jung, "Green Investment Cost Optimization Model in the Supply Chain," American Journal of Operations Research, Vol. 3 No. 6, 2013, pp. 454-462. doi: 10.4236/ajor.2013.36044.

CHAPTER 4

J. Hu, "Monotonic Vector Space Model (I): Concepts and Operations," *American Journal of Operations Research*, Vol. 4 No. 1, 2014, pp. 30-41. doi: 10.4236/ajor.2014.41004.

CHAPTER 5

J. Hu, "Monotonic Vector Space Model(II): Application in System Capability Indicator Analysis and Design,"*American Journal of Operations Research*, Vol. 4 No. 1, 2014, pp. 42-51. doi: 10.4236/ajor.2014.41005.

CHAPTER 6

Mohammad SadeghPakkar, (2014) Using DEA and AHP for Ratio Analysis. *American Journal of Operations Research*,**04**,268-279. doi: 10.4236/ajor.2014.44026

CHAPTER 7

Ambapour, S. (2015) Assessing the Relative Efficiency of Health Systems in Sub-Saharan Africa Using Data Envelopment Analysis. *American Journal of Operations Research*, **5**, 30-37. doi: 10.4236/ajor.2015.51003.

CHAPTER 8

H. Zhang, W. Song, X. Peng and X. Song, "Evaluate the Investment Efficiency by Using Data Envelopment Analysis: The Case of China," *American Journal of Operations Research*, Vol. 2 No. 2, 2012, pp. 174-182. doi: 10.4236/ajor.2012.22020.

CHAPTER 9

H. Chen, "Analysis of Studies from 2000-2010 in Real Option Theory and Application to OM," *American Journal of Operations Research*, Vol. 1 No. 1, 2011, pp. 16-23. doi: 10.4236/ajor.2011.11003.

CHAPTER 10

F. Kamoun, "Performance Analysis of Two Priority Queuing Systems in Tandem," *American Journal of Operations Research*, Vol. 2 No. 4, 2012, pp. 509-518. doi: 10.4236/ajor.2012.24060.

CHAPTER 11

C. Nwozo and C. Nkeki, "On a Dynamic Optimization Technique for Resource Allocation Problems in a Production Company," American Journal of Operations Research, Vol. 2 No. 3, 2012, pp. 357-363. doi: 10.4236/ajor.2012.23043.

CHAPTER 12

S. Suneja, S. Sharma, M. Grover and M. Kapoor, "A Different Approach to Cone-Convex Optimization," American Journal of Operations Research, Vol. 3 No. 6, 2013, pp. 536-541. doi: 10.4236/ajor.2013.36052.

CHAPTER 13

A. Kostogryzov, L. Grigoriev, G. Nistratov, A. Nistratov and V. Krylov, "Prediction and Optimization of System Quality and Risks on the Base of Modelling Processes," *American Journal of Operations Research*, Vol. 3 No. 1A, 2013, pp. 217-244. doi: 10.4236/ajor.2013.31A021.

CHAPTER 14

Delal Dink1 , Mark Langer2 , Seza Orcun3 , Joseph Pekny1 , Ronald Rardin4 , Gintaras Reklaitis1 , Behlul Saka, IMRT Optimization with Both Fractionation and Cumulative Constraints*, doi:10.4236/ajor.2011.13018.

INDEX

A

C

D

G

H

I